Maths — No Problem!

English National Curriculum

Consultant Editors
Dr Wong Khoon Yoong
Dr Julie Alderton

Series Editors
Adam Gifford
Andy Psarianos

Authors
Adam Gifford
Michelle Hayfron

MATHS NO PROBLEM!

Published by Maths — No Problem!

Printed in the United Kingdom
First Printing, 2022

ISBN 978-1-913458-54-6

Maths — No Problem!
hello@mathsnoproblem.com
www.mathsnoproblem.com

Acknowledgements
The publisher would like to thank Dr Anne Hermanson and Alex Laurie for their contributions.

Preface

Welcome to Maths — No Problem!

Maths — No Problem! is a comprehensive series of textbooks and workbooks supported by detailed online teacher guidance to deliver maths mastery in primary schools. The materials use a spiral design with carefully built-up mathematical concepts and processes adapted from the maths mastery approaches used in Singapore. An integral part of the learning process is the Concrete, Pictorial, Abstract (CPA) approach, which forms the backbone of the materials developed for this series.

Maths — No Problem! incorporates the use of concrete aids and manipulatives, problem solving, group work, individual practice and journaling.

Meet our Maths — No Problem! friends who will join you on your journey through this textbook.

The Maths — No Problem! approach is exemplified throughout the chapters by the following teaching elements:

Chapter Opener

The chapter opener depicts familiar events or contexts that serve as an introduction for pupils to the concept covered in the chapter. It can also be used as an additional prompt for journal writing or an additional exercise for pupils.

Explore

Explore is an introductory activity for pupils, which includes questions related to various lesson objectives.

Master

Each lesson contains a Master section that develops lesson objectives by introducing new concepts through a CPA approach using manipulatives and engaging pictures. Guided examples are provided to reinforce the learning objective.

Activity Time

Activity Time provides pupils with opportunities to work as individuals or in small groups to explore mathematical concepts through games and investigations.

Guided Practice

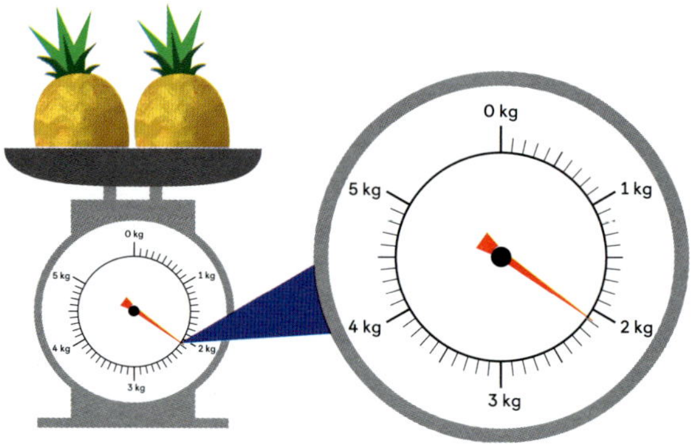

Guided Practice contains questions for further consolidation and the immediate evaluation of pupils' learning.

Mind Challenge

Mind Challenges are non-routine questions for pupils to apply relevant problem-solving skills and develop higher-order thinking skills.

carton A   carton B   carton C

My Maths Journal

My Maths Journal provides pupils with opportunities to show their understanding of the mathematical concepts learnt and to record their learning.

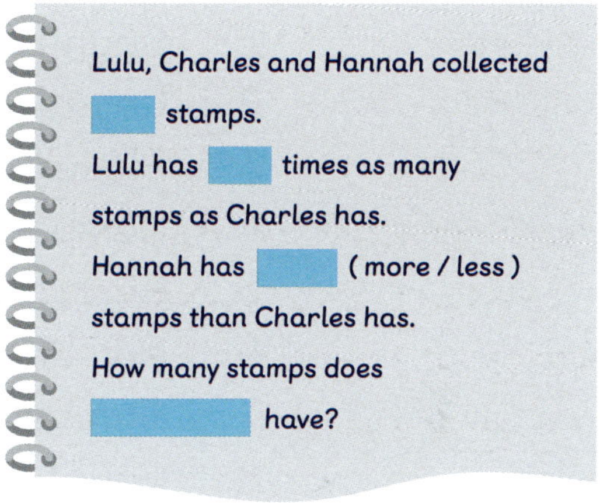

Review

Review allows pupils to reflect on their own learning after each chapter, with support from the teacher.

Review

I can...

- ☐ write an addition and subtraction fact family.
- ☐ add numbers up to 3 digits without renaming.
- ☐ add numbers up to 3 digits with renaming.
- ☐ subtract numbers up to 3 digits without renaming.
- ☐ subtract numbers up to 3 digits with renaming.
- ☐ use models to solve addition and subtraction problems, including those that require renaming.

Contents

Chapter 1 Numbers to 1000

Chapter 2 Addition and Subtraction

Page

Page

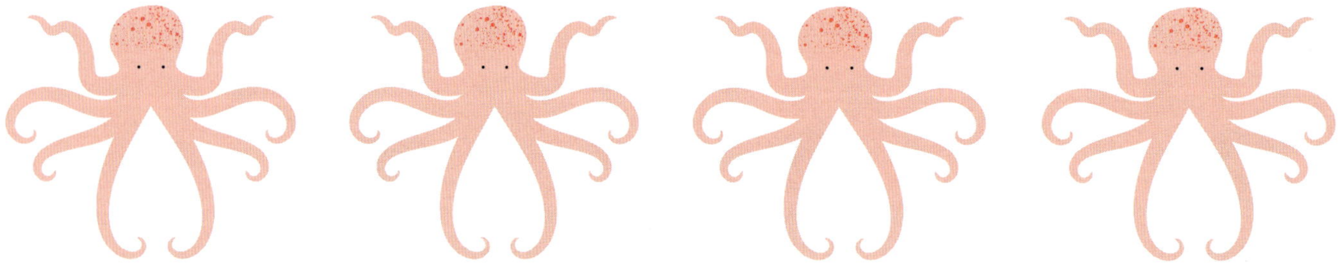

Chapter 5 Length

Page

Chapter 6 Mass

Chapter 7

Volume

Page

How many marbles are there on the shop counter in total?

Chapter 1
Numbers to 1000

Counting in Hundreds

Lesson 1

Recap

1

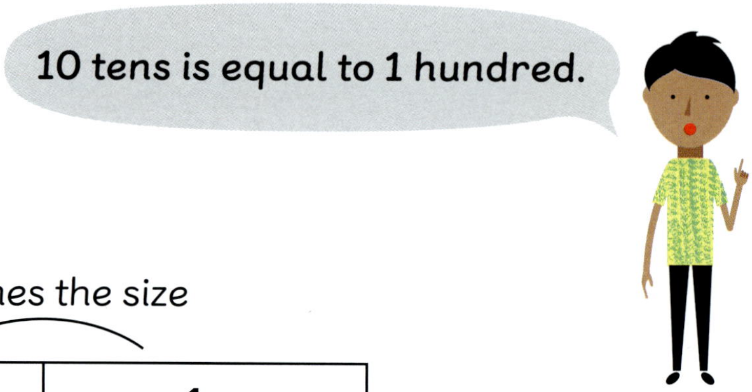

2

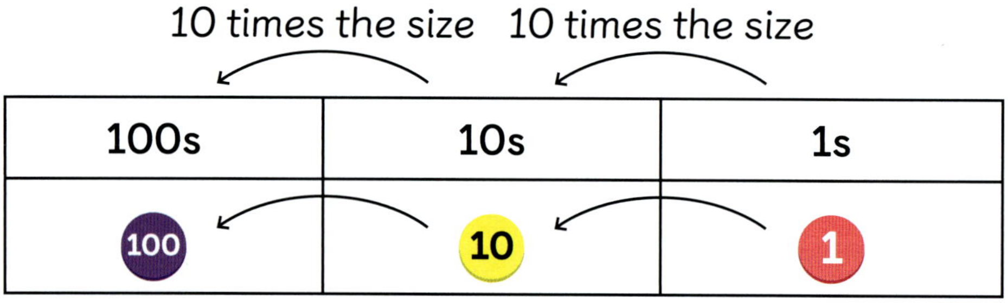

Explore

How many marbles are there?
Count in hundreds.

Master

1 Each jar contains 100 marbles.

		0	zero
		100	one hundred
		200	two hundred
		300	three hundred
	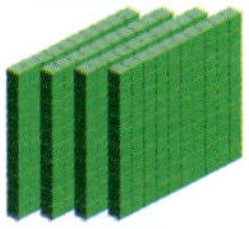	400	four hundred
		500	five hundred
		600	six hundred
		700	seven hundred
		800	eight hundred
		900	nine hundred
		1000	one thousand

2

1000 is equal to 10 hundreds.

1000								
100	200	300	400	500	600	700	800	900
10	20	30	40	50	60	70	80	90
1	2	3	4	5	6	7	8	9

Guided Practice

Count in hundreds.
Write the numbers in numerals and in words.

1

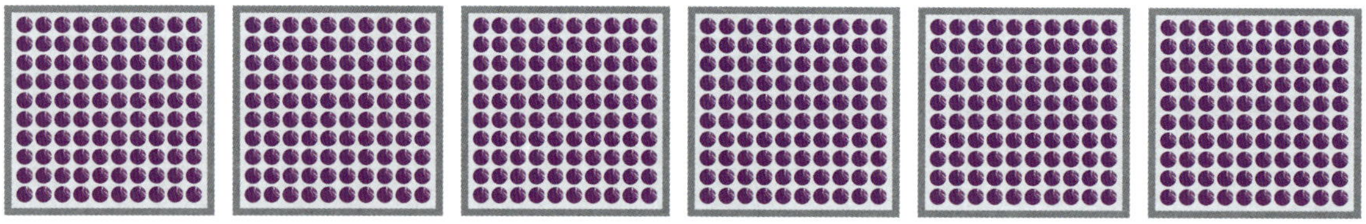

100 beads 100 beads 100 beads 100 beads 100 beads 100 beads

2

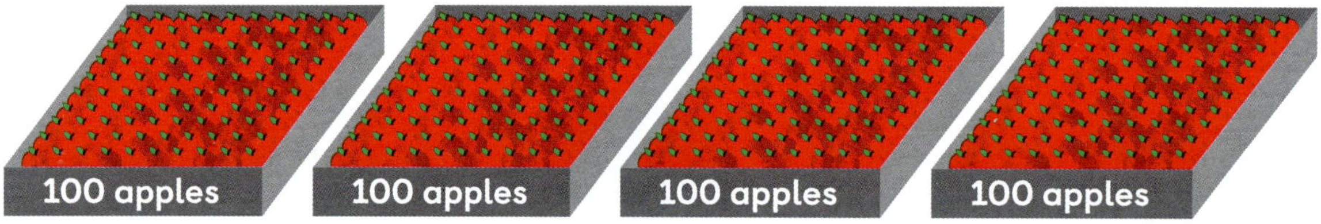

3

Complete Worksheet **1**: pages **1–4**

Counting in Hundreds, Tens and Ones

Lesson 2

Explore

How many stamps are there in total?

Master

1 Count the hundreds.

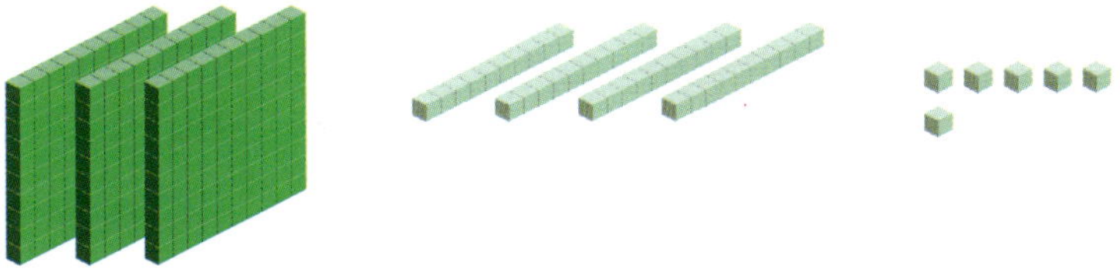

2 Count the tens.

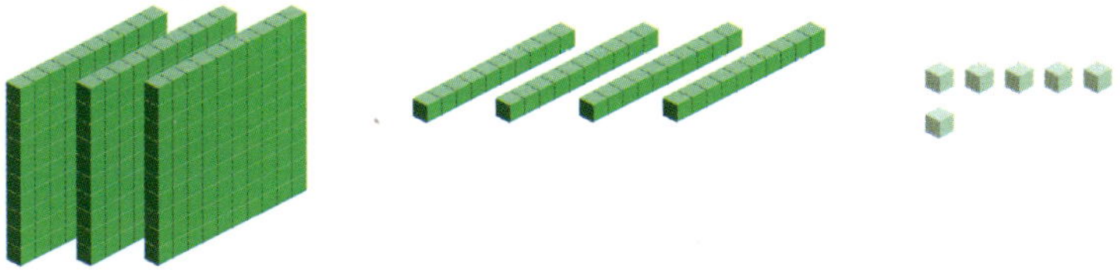

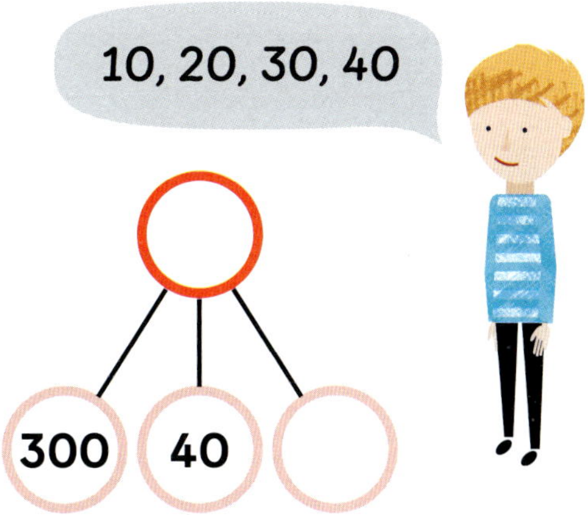

3 Count the ones.

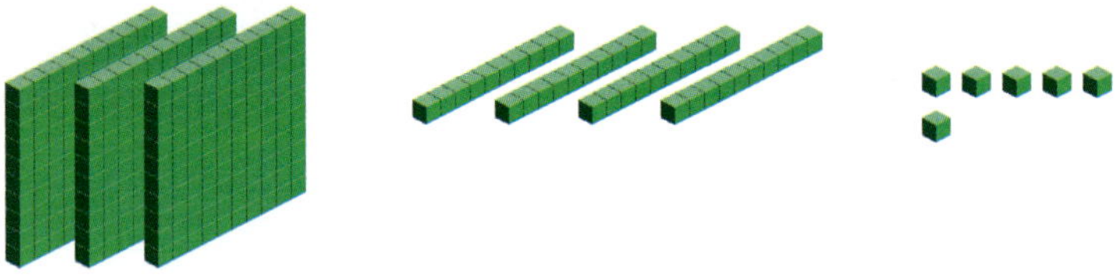

4

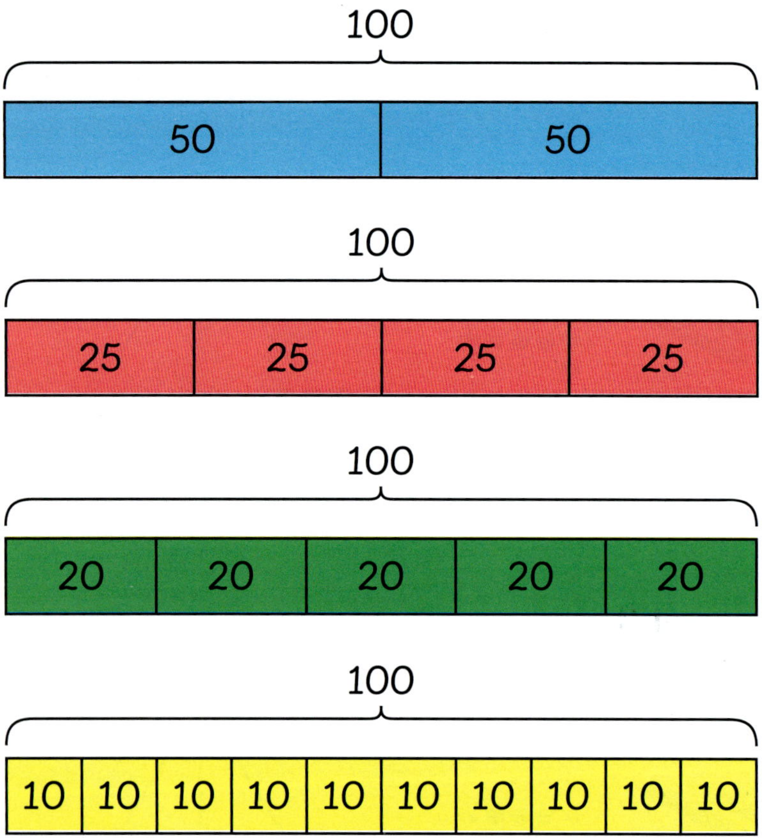

Activity Time

Partner work

You will need:

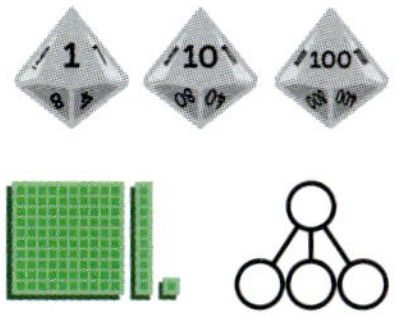

1. Roll  .
2. Ask your partner to show the numbers on the dice using 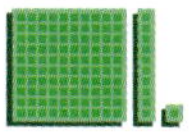 and 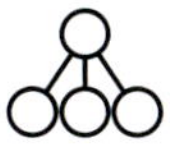 .
3. Work together to complete four 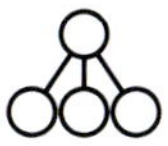, showing your number in different ways.

For example:

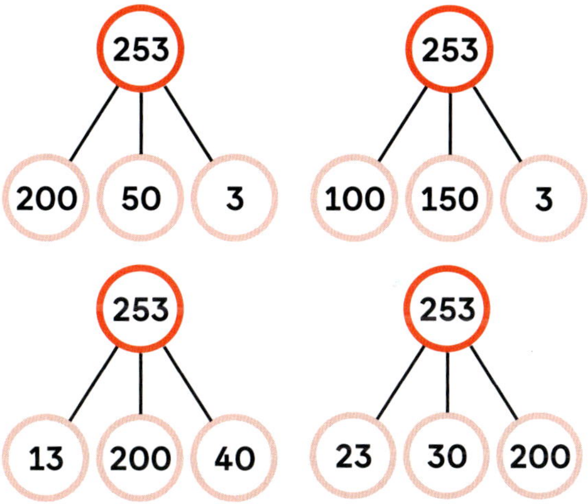

4. Check the answers are correct.

Guided Practice

Count and write the number.
Then complete the number bond.

1

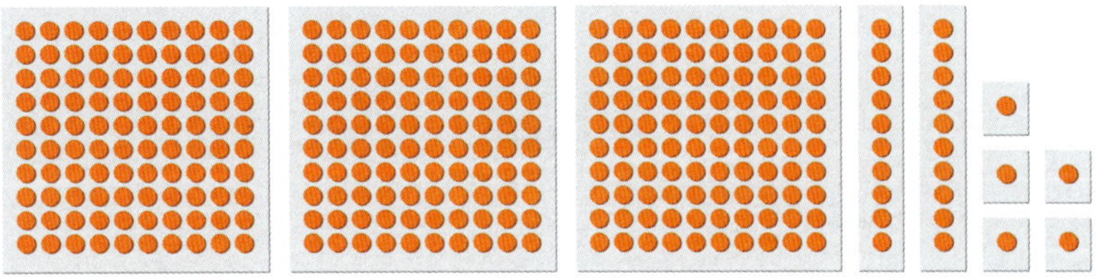

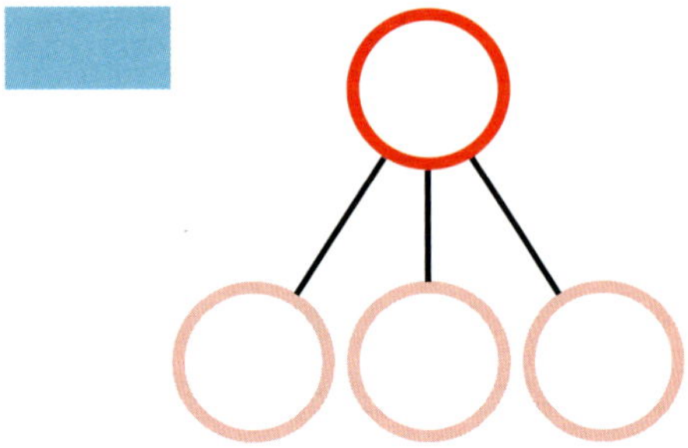

2

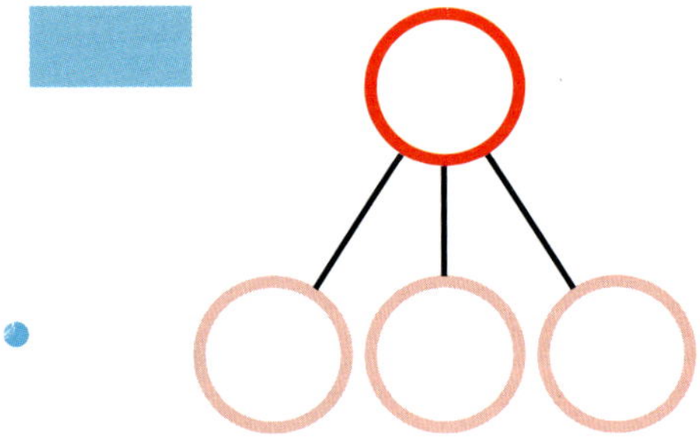

3

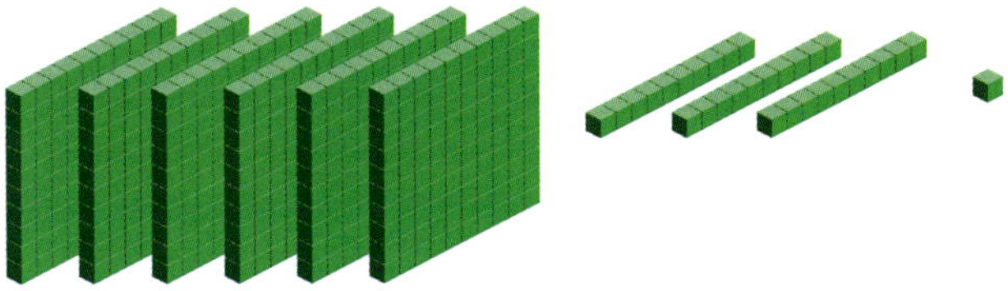

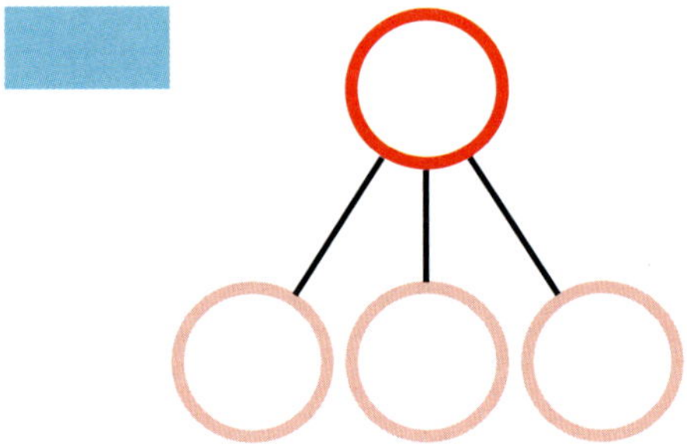

Complete Worksheet **2**: pages **5–6**

Place Value

Lesson 3

Explore

What do each of the digits in 362 stand for?

Master

1

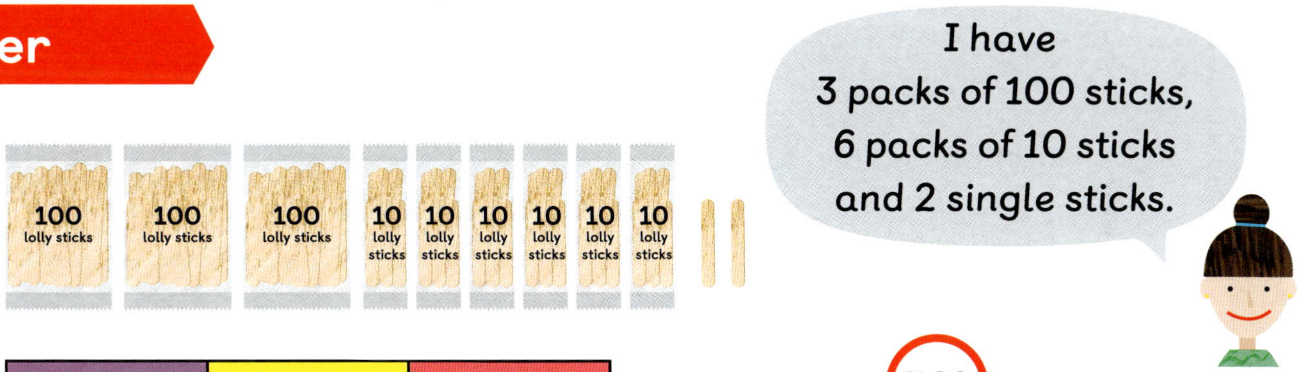

hundreds	tens	ones
3	6	2

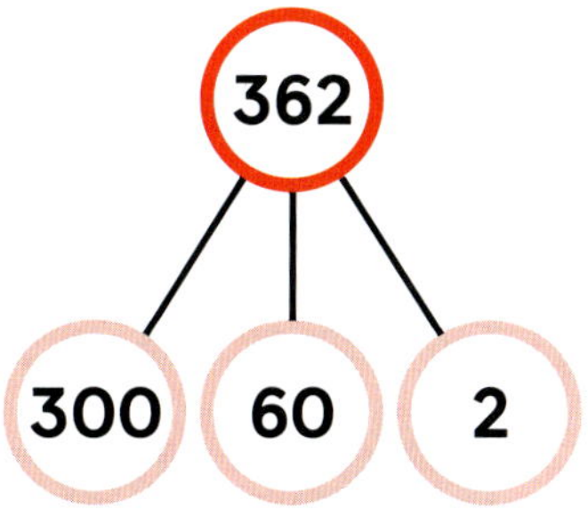

362 = 3 hundreds + 6 tens + 2 ones
362 = 300 + 60 + 2

The digit 3 stands for 300.
The digit 6 stands for 60.
The digit 2 stands for 2.

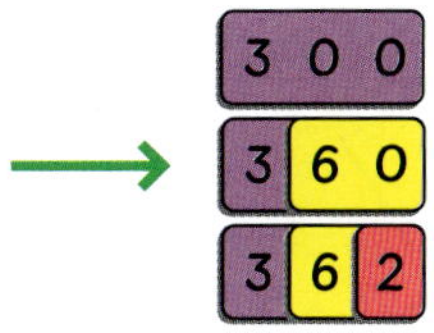

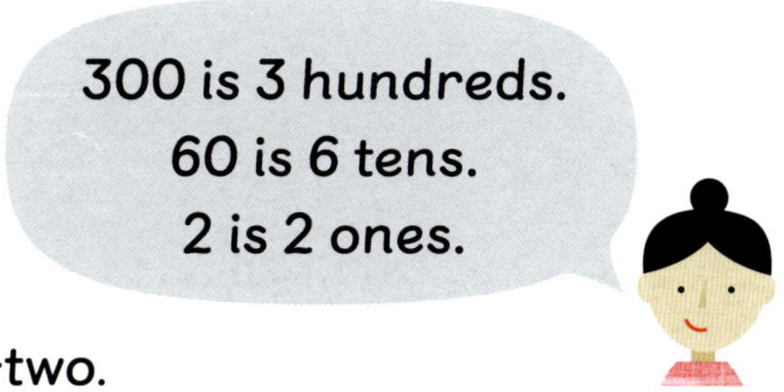

362 is written as **three hundred and sixty-two.**

2 Can you find the value of each digit in 470?

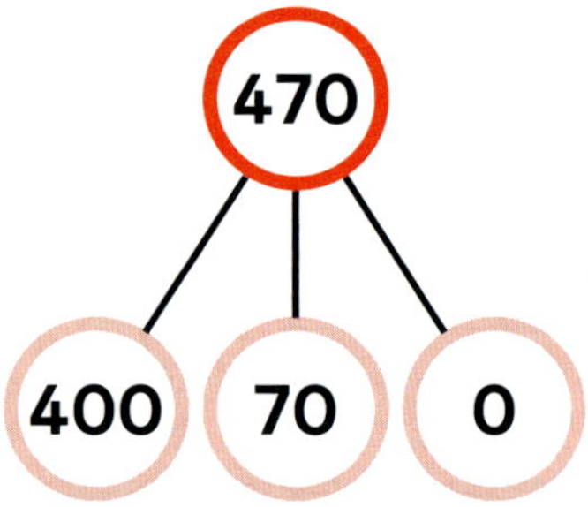

470 = ☐ hundreds + ☐ tens + ☐ ones

470 = ☐ + ☐ + ☐

The value of the digit 4 is 4 0 0.
The value of the digit 7 is 7 0.
The value of the digit 0 is 0.

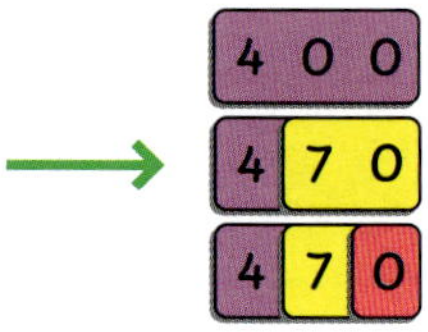

470 is written as **four hundred and seventy.**

Activity Time

Group work (3 pupils)

You will need:

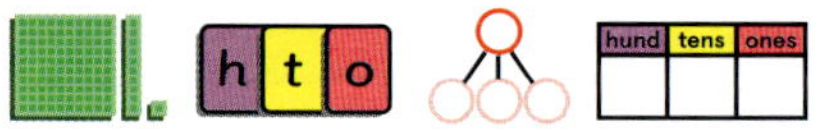

(1) Pupil A shows a 3-digit number using 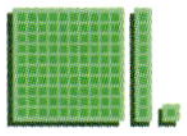 .

(2) Pupil B shows the number using h t o.

(3) Pupil C writes the number on a 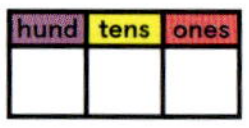 .

(4) Work together to show the 3-digit number in different ways using 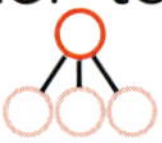 .

(5) Switch roles and repeat the activity.

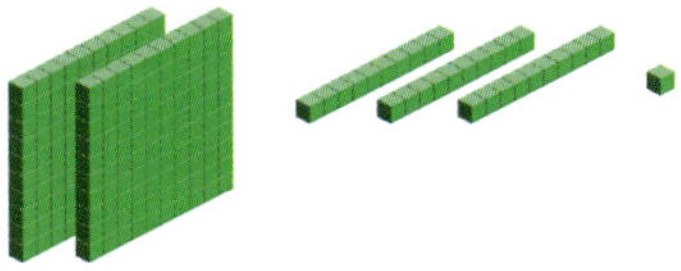

hundreds	tens	ones
2	3	1

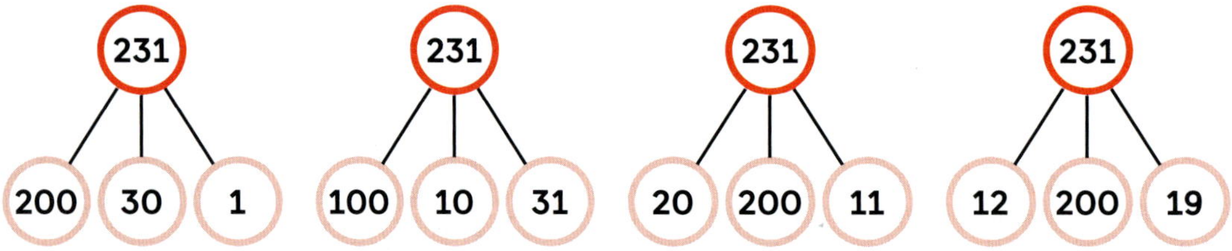

Guided Practice

1 Look at the place-value chart and fill in the blanks.

(a)

hundreds	tens	ones
3	7	9

In 379, the digit 9 stands for ____ .

In 379, the digit 3 stands for ____ .

(b)

hundreds	tens	ones
5	0	6

In 506, the digit 0 stands for ____ .

In 506, the digit 5 stands for ____ .

2 Use 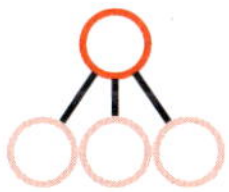 to show each number in two different ways.

(a) 192

(b) 250

(c) 705

Complete Worksheet 3: pages 7–8

Comparing and Ordering Numbers

Lesson 4

Explore

Who has more stickers, Sam or Lulu?

Master

1 Compare 253 and 326. Which number is greater?

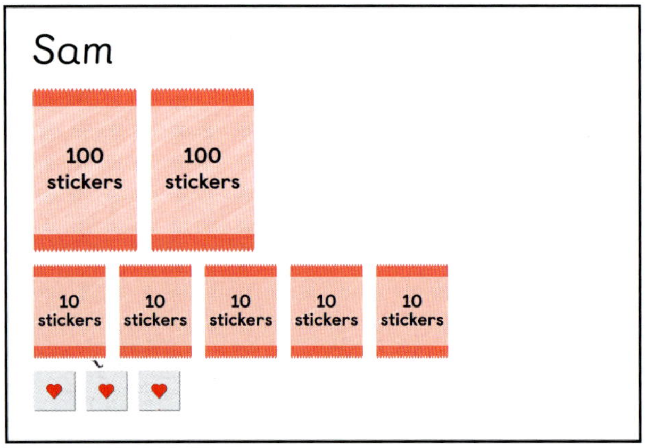

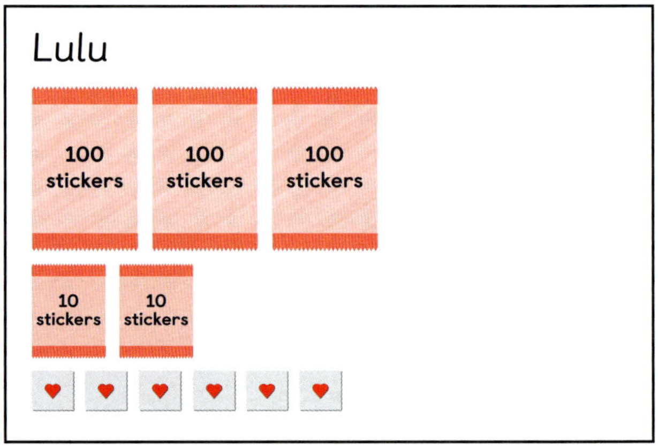

hundreds	tens	ones
2	5	3

hundreds	tens	ones
3	2	6

3 hundreds is greater than 2 hundreds.
326 is greater than 253.
Lulu has more stickers.

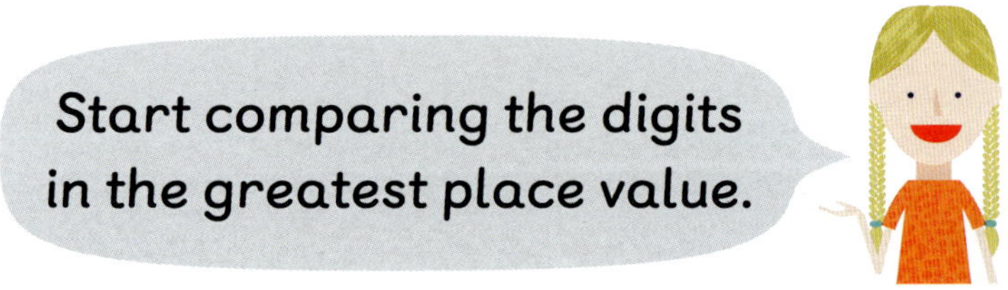

2 Compare 764 and 758. Which number is smaller?

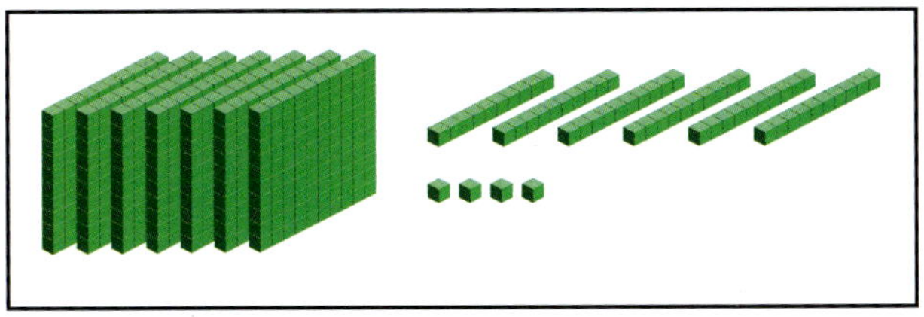

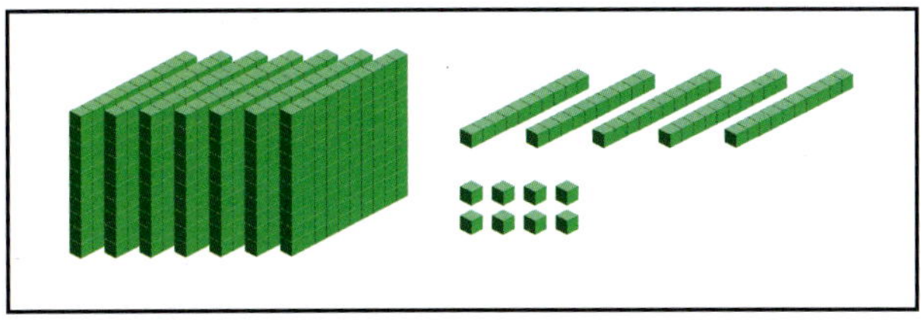

hundreds	tens	ones
7	6	4

hundreds	tens	ones
7	5	8

First compare the hundreds. They are the same, so we compare the tens.

5 tens is smaller than 6 tens. 758 is smaller than 764.

758 is smaller than 764.

3 Put the numbers in order, starting with the smallest.

| 4 | 9 | 5 | | 5 | 4 | 9 | | 5 | 9 | 4 |

5 hundreds is greater than 4 hundreds.

____ is the smallest number.

Now, compare tens of 549 and 594.

____ tens is less than ____ tens.

____ is smaller than ____.

____ is the greatest number.

594 > 549

____, ____, ____

smallest → greatest

4 Estimate and place the following numbers on the number line.
Put the numbers in order, starting with the greatest.

300 800 260 720 543 121

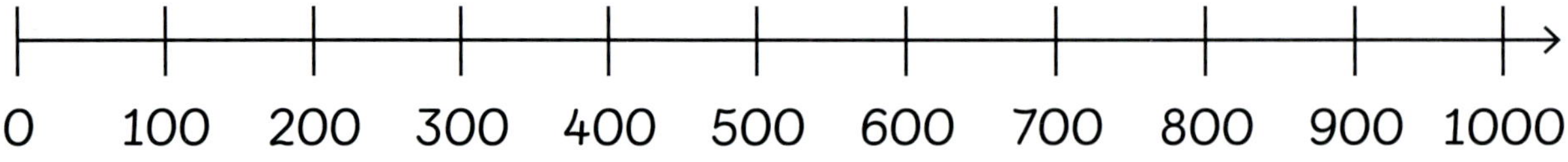

___ , ___ , ___ , ___ , ___ , ___

greatest → smallest

Activity Time

Play a game (3 pupils)

You will need: number cards 0–9, h t o place value counters, hund tens ones place value chart

1. Take turns to pick three number cards each.
2. Write each digit you pick on the hund | tens | ones chart.
3. Show your number using h t o.
4. Compare your numbers.
5. The pupil with the greatest number wins.

Guided Practice

1 Compare the numbers. Use **more than**, **less than** or **equal to**.

(a)

69 96

69 is ______ 96.

(b)

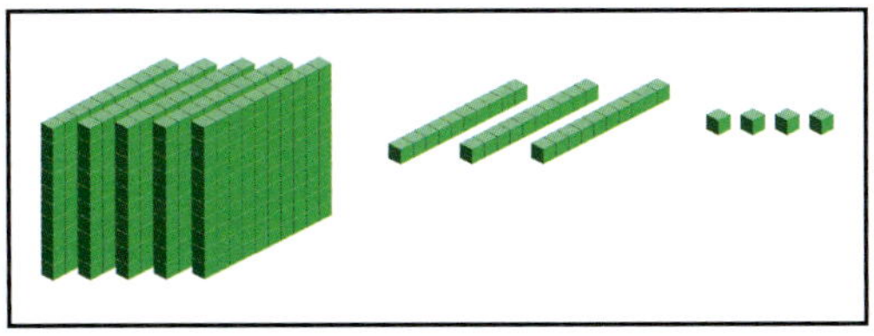

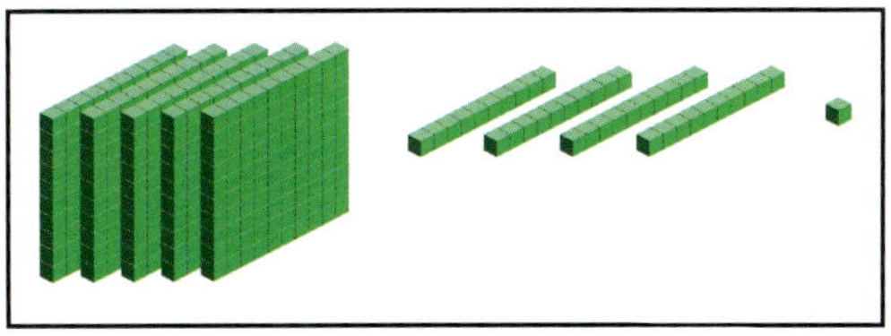

543 is ______ 534.

(c) 629 is ______ 629.

(d) 354 is ______ 349.

2 Put the numbers in order.

(a) Begin with the smallest.

(i) 53 35 56

______, ______, ______

(ii) 284 728 287

______, ______, ______

(b) Begin with the greatest.

(i) 68 85 63

______, ______, ______

(ii) 560 605 650

______, ______, ______

Complete Worksheet **4**: pages **9–10**

Counting in Fifties

Lesson 5

Explore

Each stack has 50 cards.
How many cards are there on the table?

Master

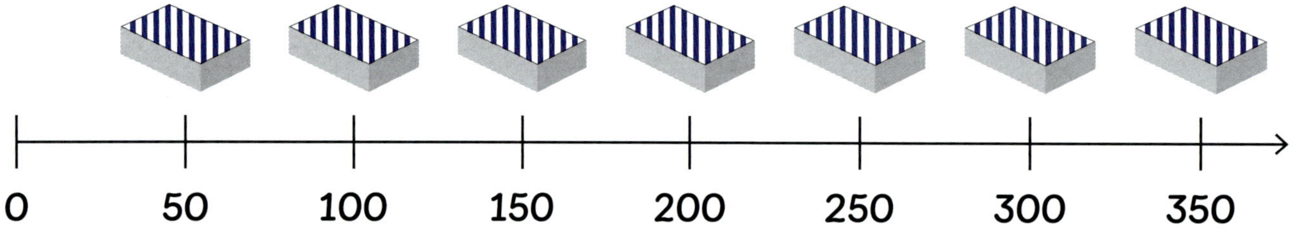

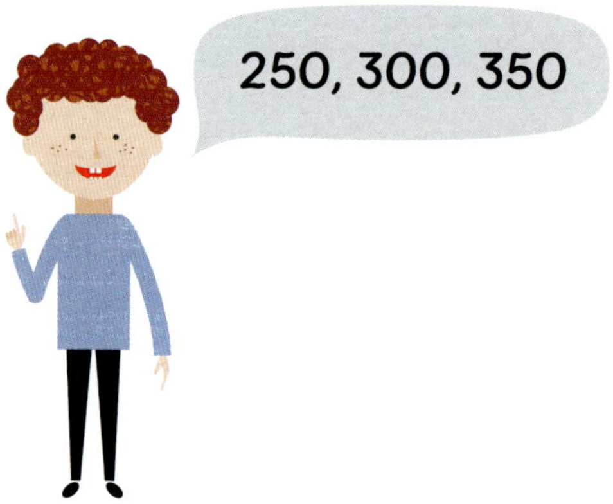

There are 350 cards on the table.

Activity Time

Play a game (2 pupils)

You will need:

1. Pick a card each.
2. The number on the card shows the number of fifties you need to count.
3. Take turns to count on in fifties.
4. Write the number you reach on a whiteboard. This is the number you will start from when you pick your next card.
5. Keep taking cards until you have counted to 1000.
6. The pupil who reaches 1000 first wins.
7. Repeat the activity.

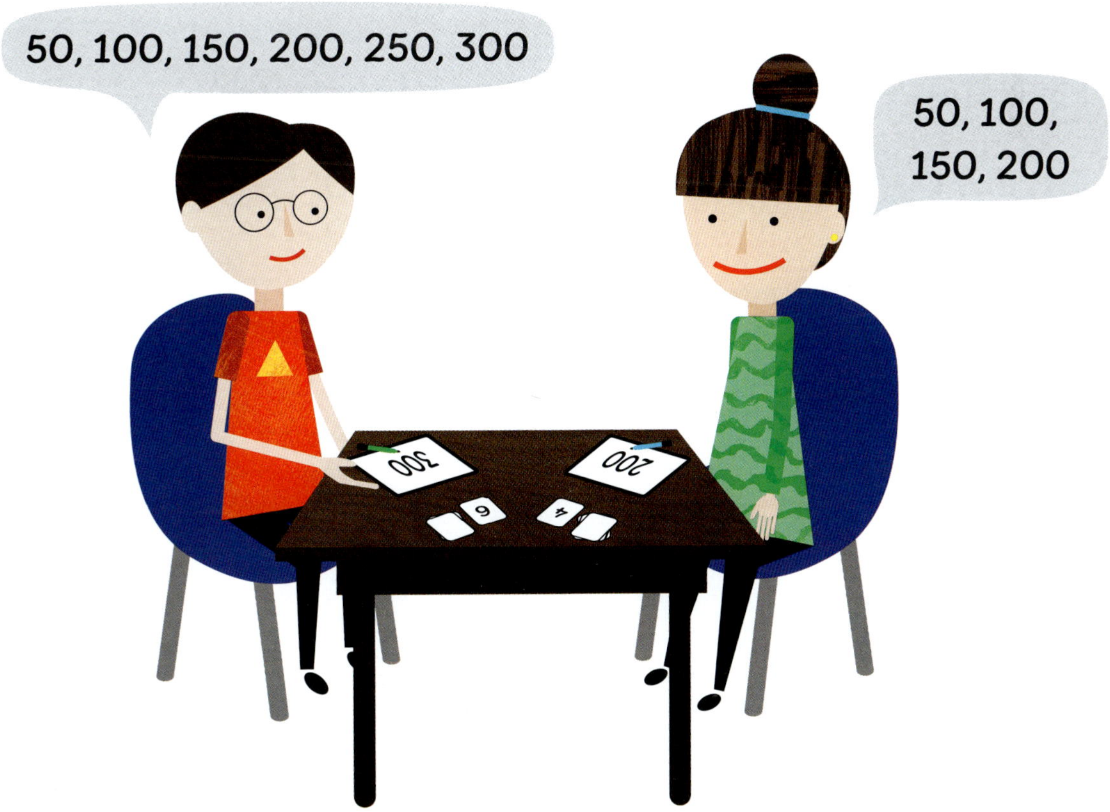

Guided Practice

1 Count, then write the number.

(a) 

Each plate has 5 hot dogs. ____

(b) 

Each bag has 50 counters. ____

2 Fill in the blanks to complete the number patterns.

(a) 50, ____, 150, 200, ____, 300, 350, ____, 450, 500

(b) 250, 300, ____, 400, 450, ____, ____, 600, 650, ____

Complete Worksheet **5**: pages **11–12**

Number Patterns

Lesson 6

Explore

Can you help Lulu find the missing number?

Master

1 What is 1 more than 316?

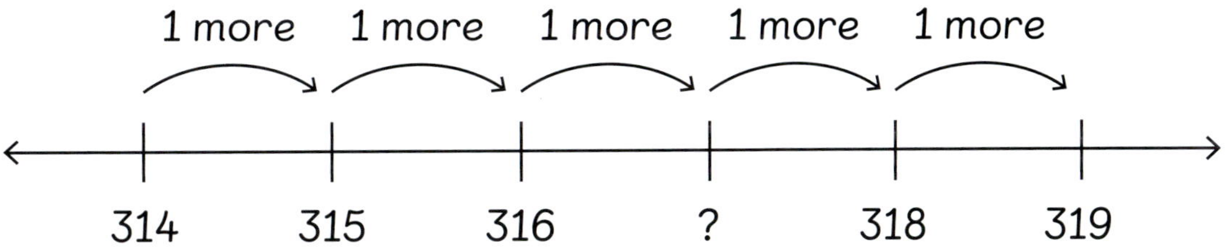

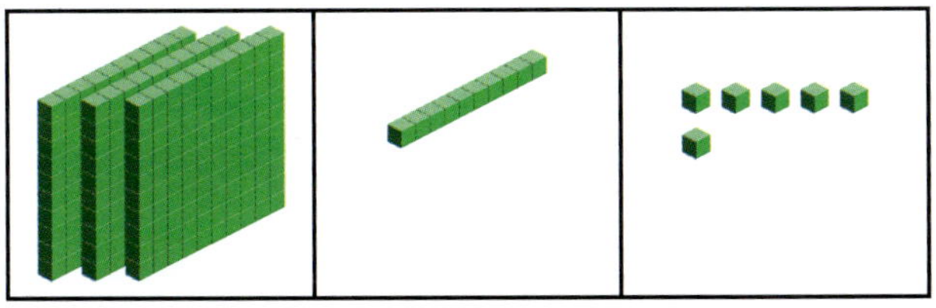

hundreds	tens	ones
3	1	6

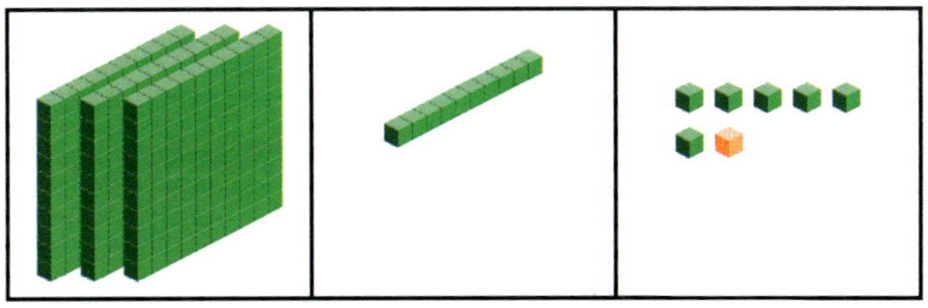

hundreds	tens	ones
3	1	7

1 more than 316 is 317.

2 What is 10 more than 334?

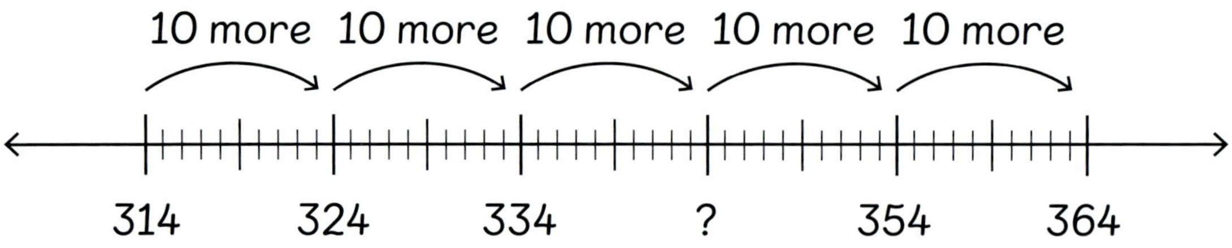

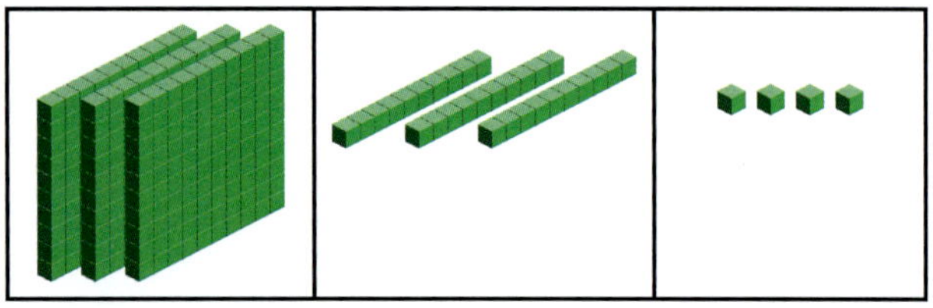

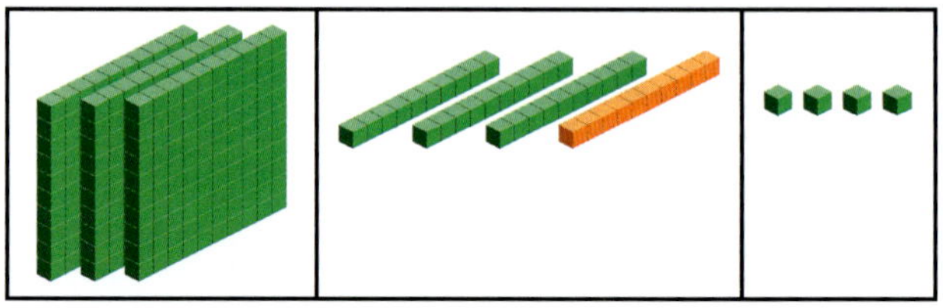

hundreds	tens	ones
3	3	4

hundreds	tens	ones
3	4	4

10 more than 334 is [ ].

[ ] is the missing number.

3 What is 10 less than 294?

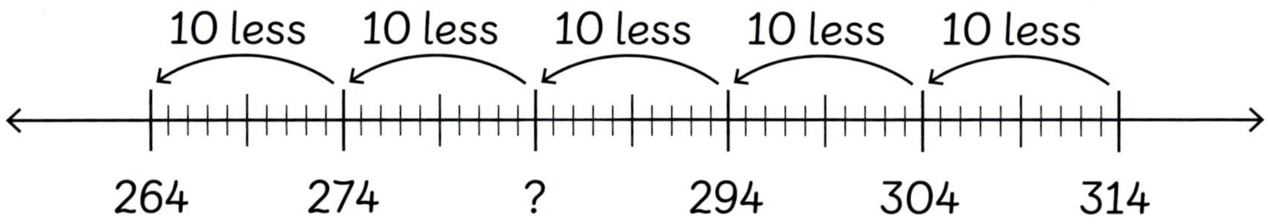

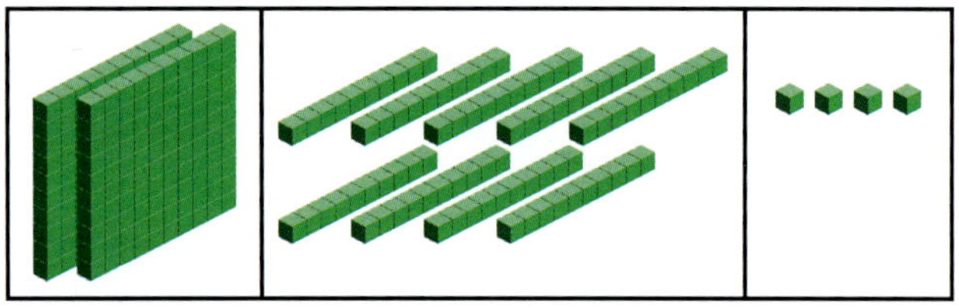

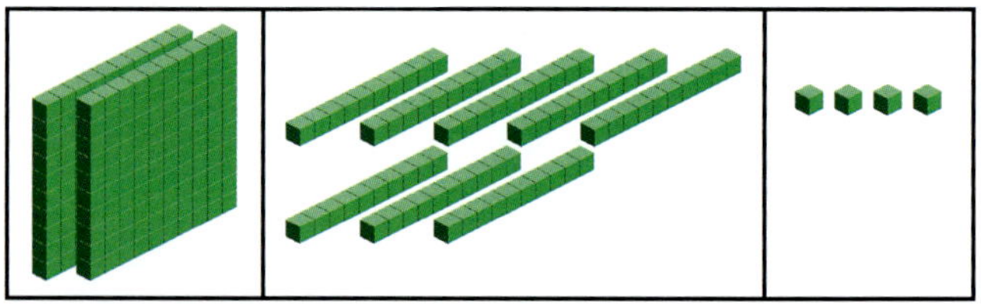

hundreds	tens	ones
2	9	4

hundreds	tens	ones
2	[ ]	4

10 less than 294 is [ ].

[ ] is the missing number.

Guided Practice

1 Fill in the blanks.

(a)

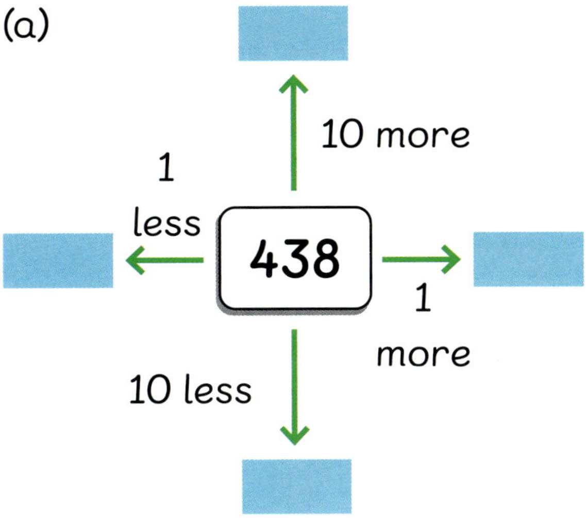

(b)

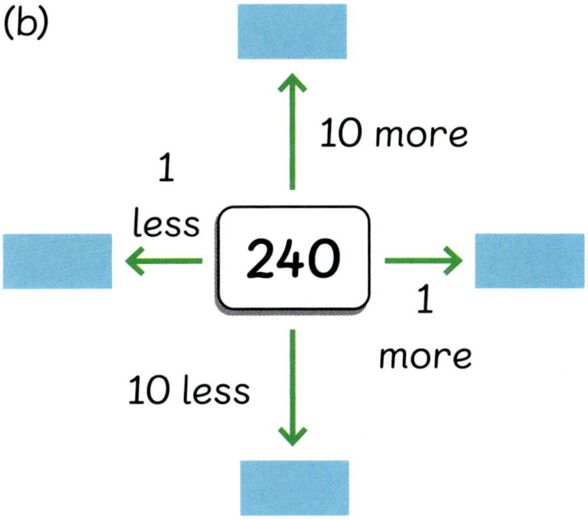

2 Find the missing numbers.

(a) 1 more than 679 is ____.

(b) 1 less than 460 is ____.

(c) 10 more than 485 is ____.

(d) 10 less than 459 is ____.

3 Fill in the blanks to complete the number patterns.

(a) 436, 437, ____, 439, ____, ____, 442

(b) 436, 446, ____, ____, 476, ____, 496

(c) 436, 426, 416, ____, ____, 386, ____, 366

Complete Worksheet **6**: pages **13–14**

Number Patterns

Lesson 7

Explore

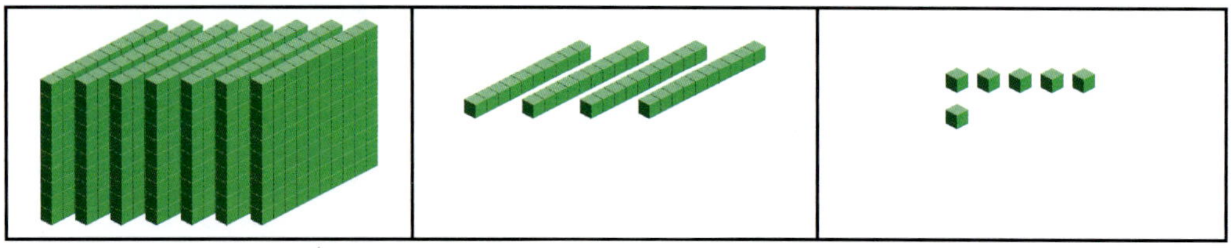

hundreds	tens	ones
7	4	6

Jacob is working on number patterns.
Can you help him find what is 100 more than 746?

Master

1

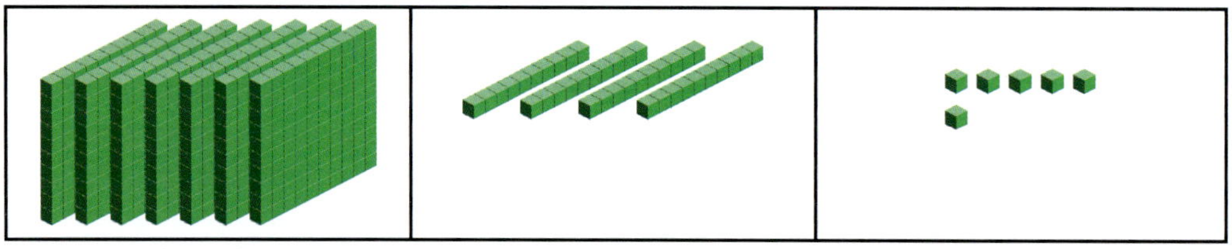

hundreds	tens	ones
7	4	6

100 more ↓

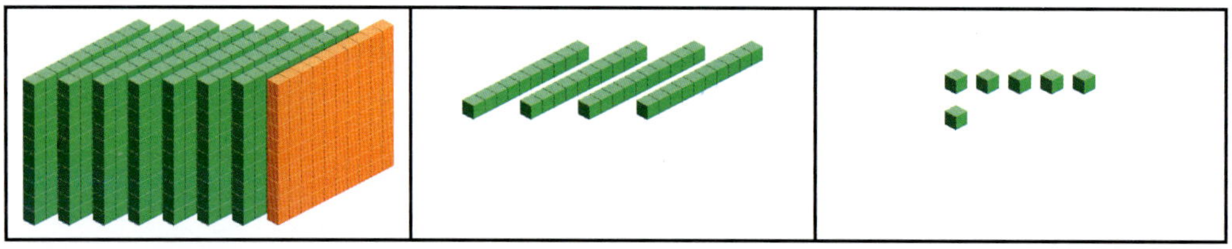

hundreds	tens	ones
8	4	6

100 more than 746 is 846.

2 What is 100 less than 746?

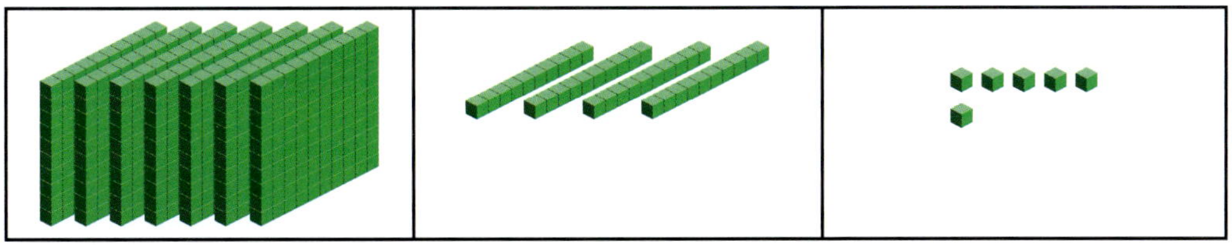

hundreds	tens	ones
7	4	6

100 less ↓

hundreds	tens	ones
6	4	6

100 less than 746 is 646.

3

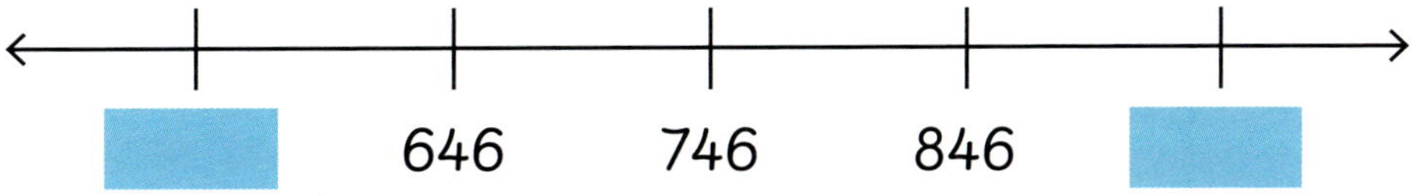

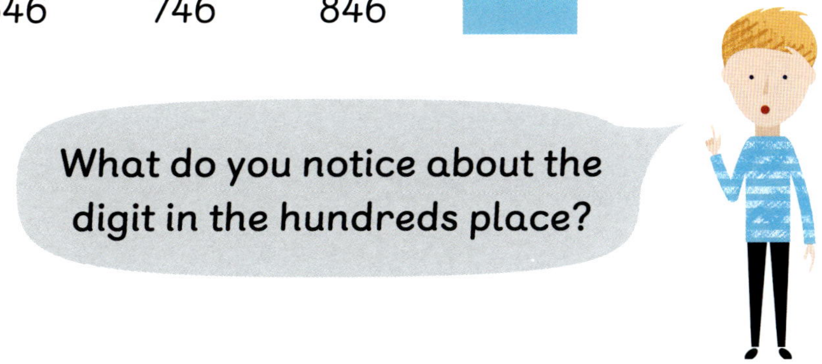

Guided Practice

1 Find the missing numbers.

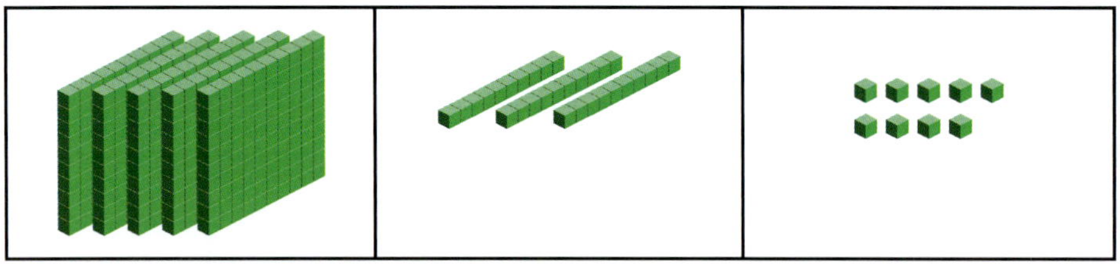

hundreds	tens	ones
5	3	9

(a) 100 more than 539 is ____.

(b) 100 less than 539 is ____.

2 Complete the table.

	100 less		100 more
(a)	____	126	____
(b)	537	____	____
(c)	____	____	625

3 Fill in the blanks to complete the number patterns.

(a) 271, 371, ____, ____, 671

(b) 893, 793, ____, ____, 493

(c) ____, ____, 385, 485, 585

Complete Worksheet 7: pages 15–16

Counting in Fours and Eights

Lesson 8

Explore

How many pencils are there altogether?

Master

1 5 packs of 10 pencils is 50 pencils.

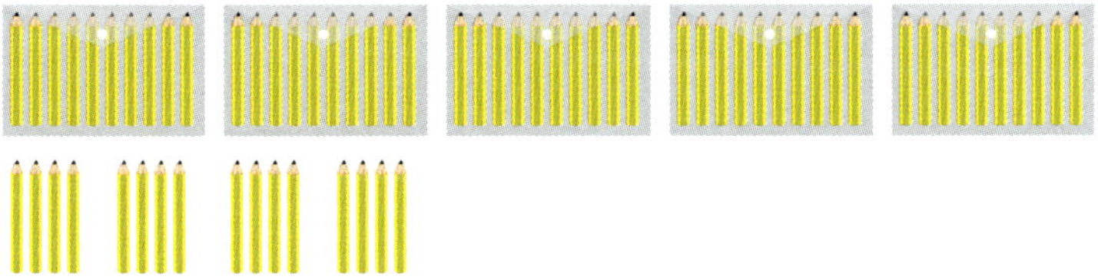

There are 66 pencils altogether.

2

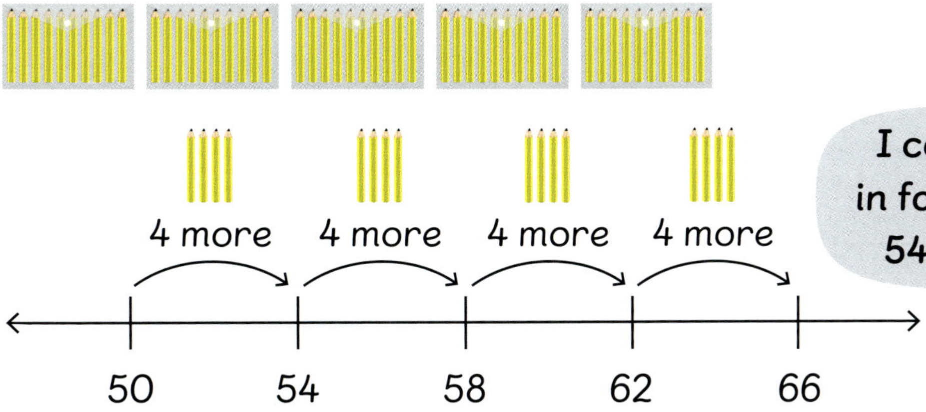

3

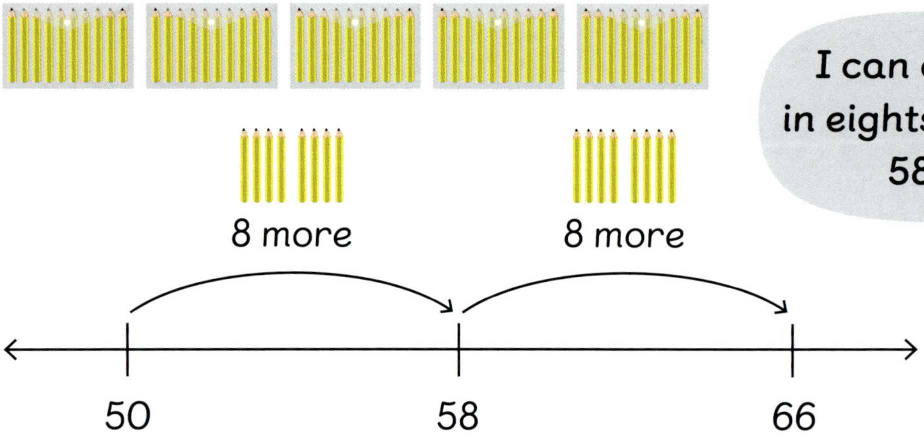

Guided Practice

1 Count the pears in fours.
Write the number.

2 Fill in the blanks.

(a)

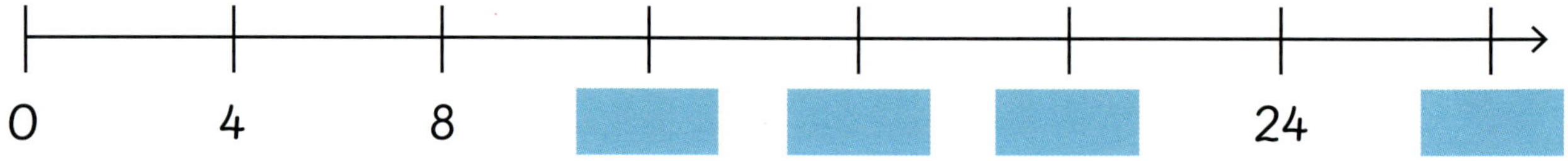

(b)

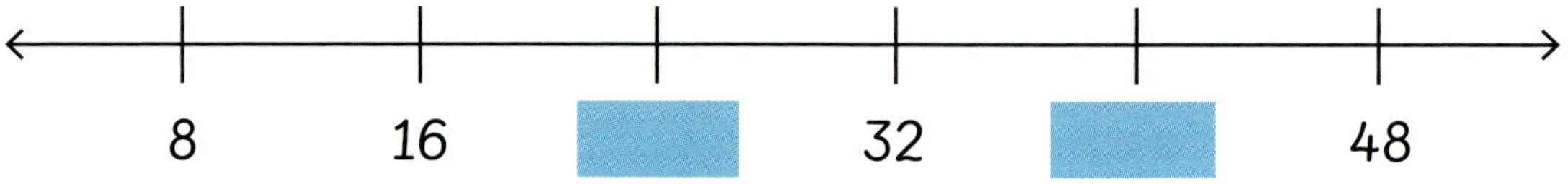

Complete Worksheet **8**: pages **17–18**

Mind Challenge

Oak is thinking of a number.

> 100 more than my number
> is equal to 10 less than 630.
> What is my number?

My Maths Journal

Use the digit cards to make the smallest 3-digit number possible.

4	8	5	2

Explain why you think this is the smallest number.

Review

I can...

- ☐ count from 0 to 1000.
- ☐ count in hundreds, tens and ones.
- ☐ tell the value of a digit in a 3-digit number.
- ☐ compare and order numbers from 0 to 1000.
- ☐ count in fifties.
- ☐ find 1, 10 and 100 more than or less than a number.
- ☐ complete number patterns.
- ☐ count in fours and eights.

How many stickers does Ruby have in total?

Chapter 2
Addition and Subtraction

Addition and Subtraction Facts

Lesson 1

Explore

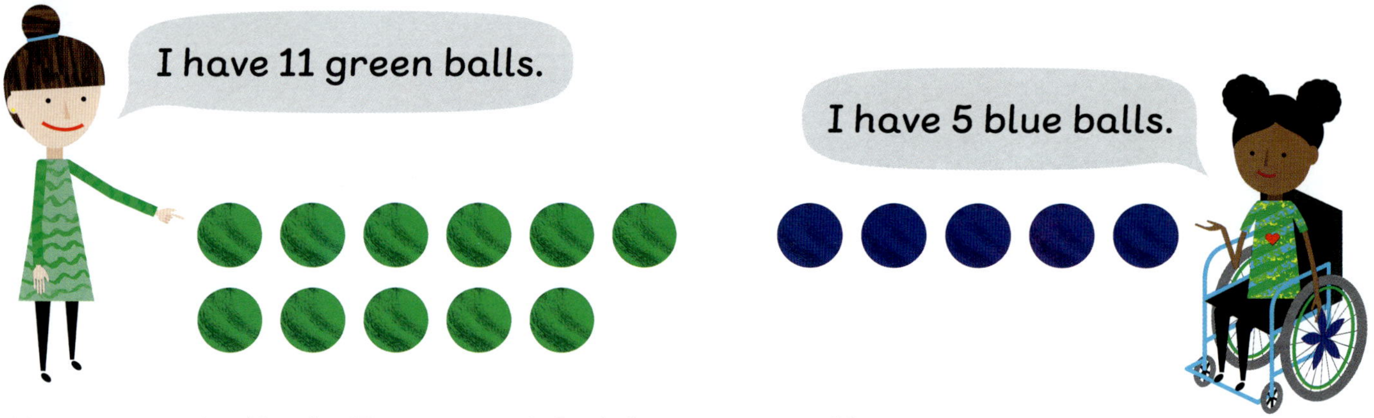

How many balls do Emma and Oak have in total?

Master

1 Add to find the total.

11 green balls and 5 blue balls make 16 balls.
11 + 5 = 16

5 blue balls and 11 green balls make 16 balls.
5 + 11 = 16

Emma and Oak have 16 balls in total.

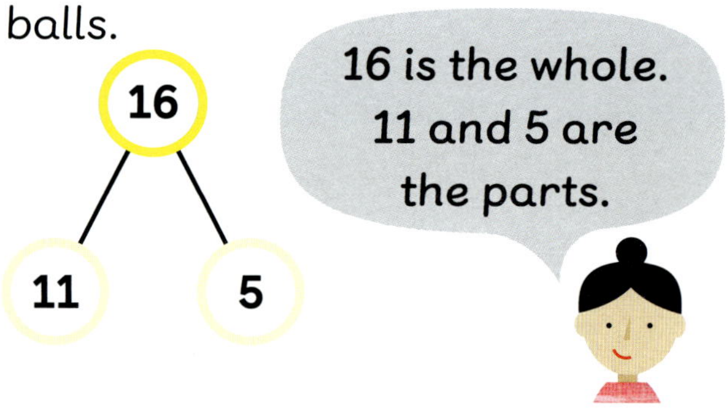

2 Subtract to find the parts.

Subtract 11 green balls from 16 balls to find the number of blue balls.
16 − 11 = 5

Subtract 5 blue balls from 16 balls to find the number of green balls.
16 − 5 = 11

3

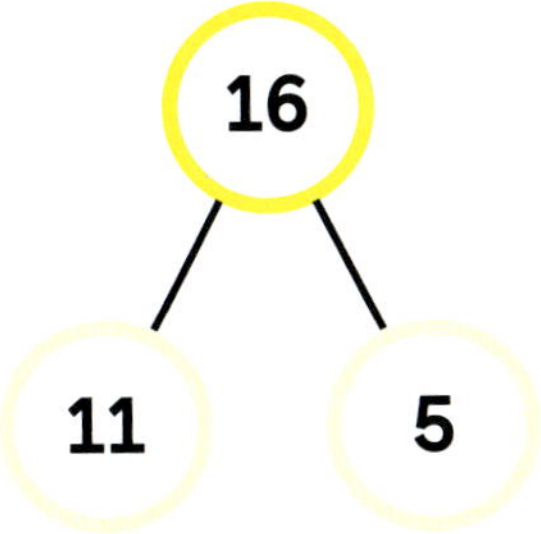

This is an addition and subtraction fact family.

11 + 5 = 16	16 − 11 = 5
5 + 11 = 16	16 − 5 = 11

4

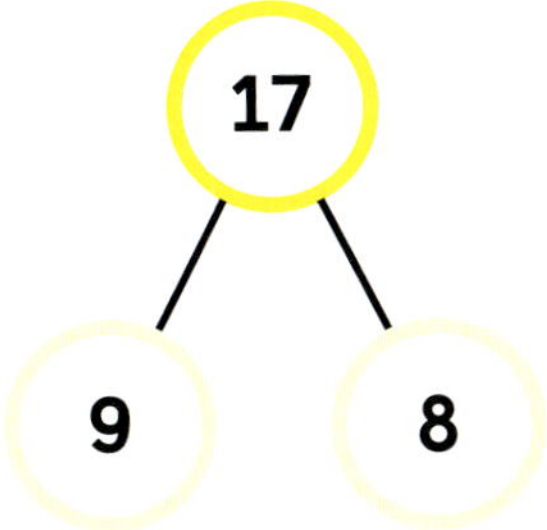

9 + 8 = 17	17 − 9 = 8
8 + 9 = 17	17 − 8 = 9

17 is the whole.
8 and 9 are the parts.

Activity Time

Partner work

You will need:

1. Pick two cards.
2. Use the numbers on the cards to write two addition equations.
3. Ask your partner to write two subtraction equations using the same three numbers in your addition equations.
4. Tell your partner a number story using the addition and subtraction fact family.
5. Repeat the activity, taking turns.

There are 11 apples on the tree. I take 3 apples. Now there are 8 apples on the tree.

Guided Practice

Complete each addition and subtraction fact family.

1 $12 + 6 = 18$

___ + ___ = ___

___ − ___ = ___

___ − ___ = ___

2 $3 + 16 = 19$

___ + ___ = ___

___ − ___ = ___

___ − ___ = ___

3 $18 - 7 = 11$

___ − ___ = ___

___ + ___ = ___

___ + ___ = ___

4 $15 - 6 = 9$

___ − ___ = ___

___ + ___ = ___

___ + ___ = ___

Complete Worksheet **1**: pages **25–26**

Adding Ones

Lesson 2

Explore

314 children were sitting in the school hall.
3 more children came into the school hall.
How many children were in the school hall altogether?

Master

314 + 3 = ▭

Method 1 Count on from 314.

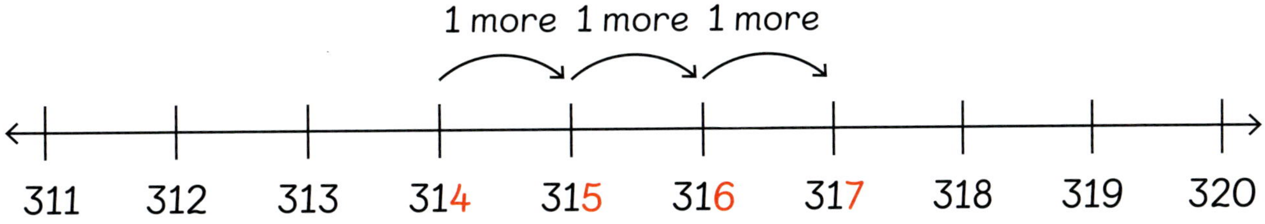

314 + 3 = 317

Method 2 Add the ones.

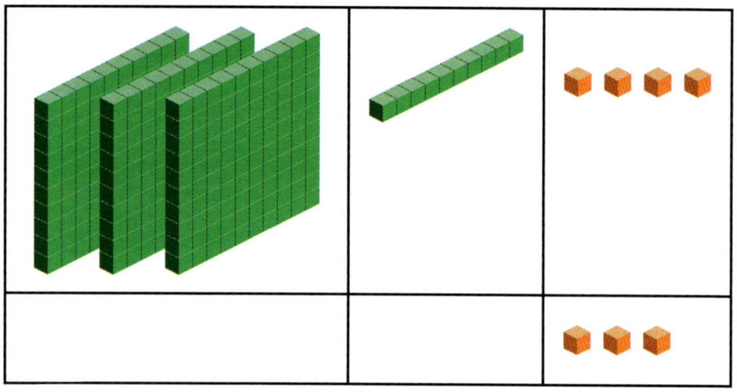

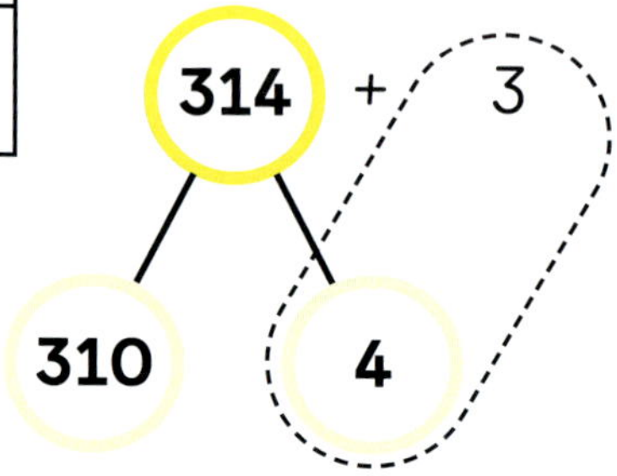

h	t	o
3	1	4
+		3
3	1	7

4 + 3 = 7
310 + 7 = 317

314 + 3 = 317

There were 317 children in the school hall altogether.

Guided Practice

Add.

1 132 + 5 = ▢

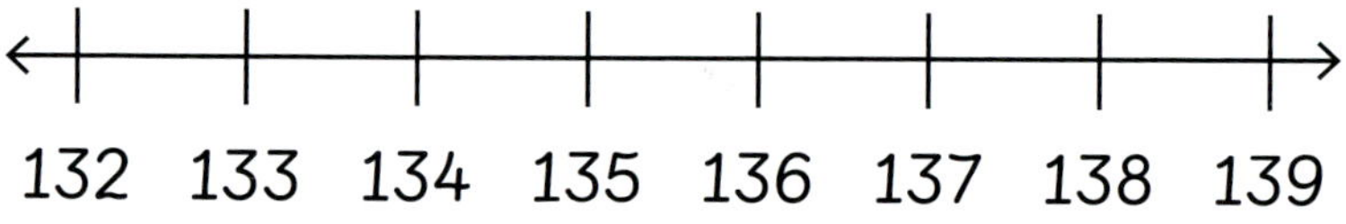

2 3 + 215= ▢

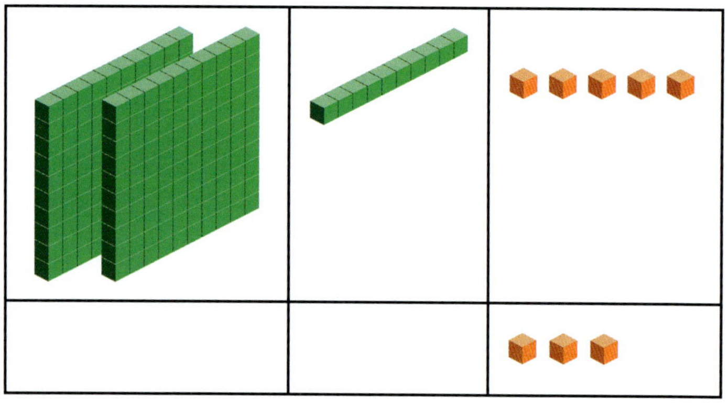

3 473 + 6 = ▢

4 8 + 560 = ▢

Complete Worksheet **2**: pages **27–28**

Adding Tens

Lesson 3

Explore

There are 325 books in the class library.
Lulu places 70 more books in the class library.
How many books are there in the class library in total?

Master

1 325 + 70 = ▭

Count on in tens from 325.

Method 1

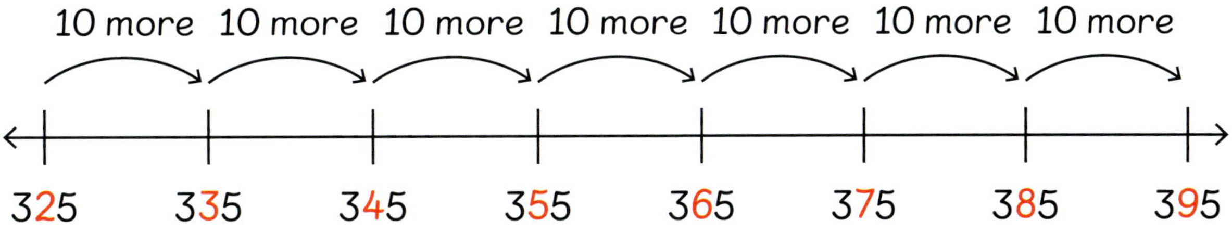

325 + 70 = 395

Method 2 Add the tens.

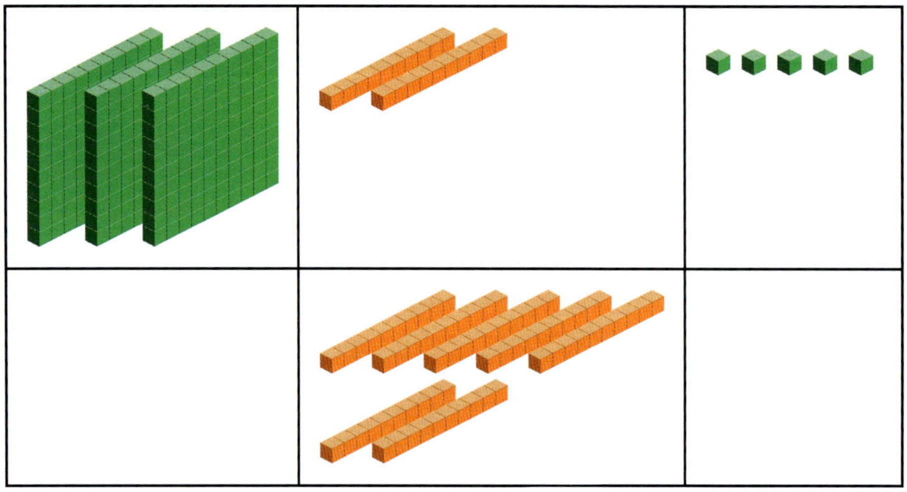

	h	t	o
	3	2	5
+		7	0
	3	9	5

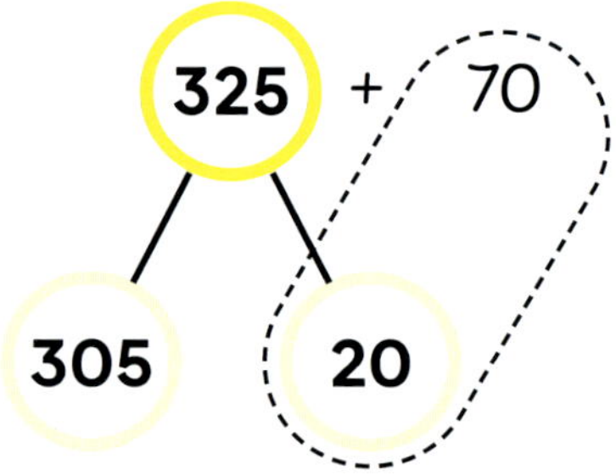

20 + 70 = 90
305 + 90 = 395

325 + 70 = 395

There are 395 books in the class library in total.

2 416 + 30 =

Method 1

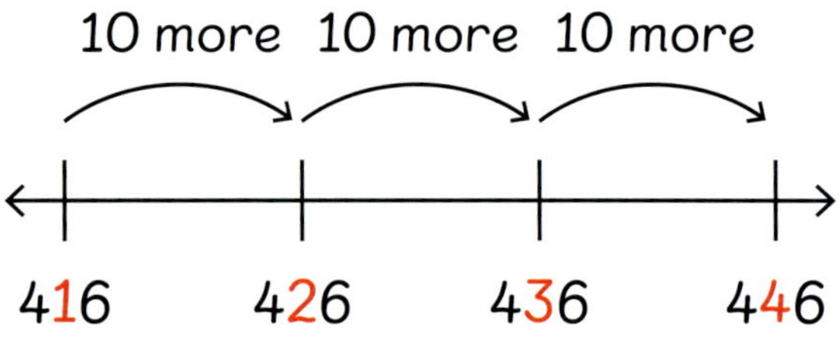

416 + 30 = 446

Method 2 Add the tens.

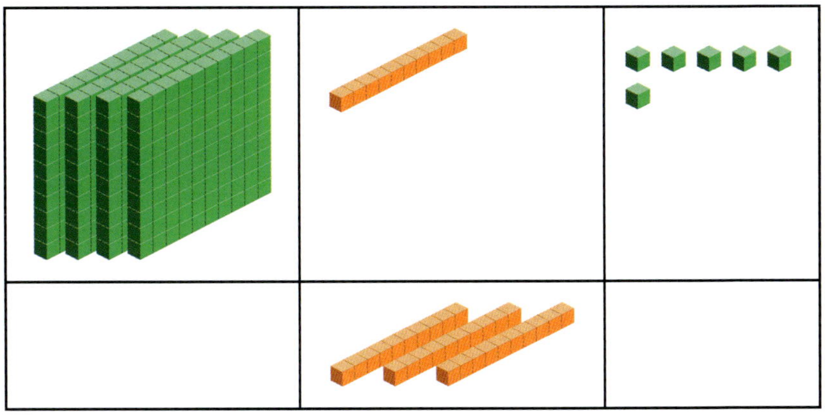

	h	t	o
	4	1	6
+		3	0
	4	4	6

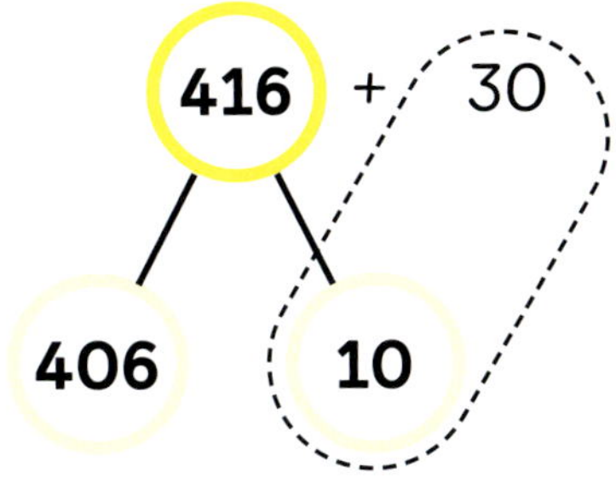

10 + 30 = 40

406 + 40 = 446

416 + 30 = 446

Guided Practice

Add.

1 623 + 50 = ______

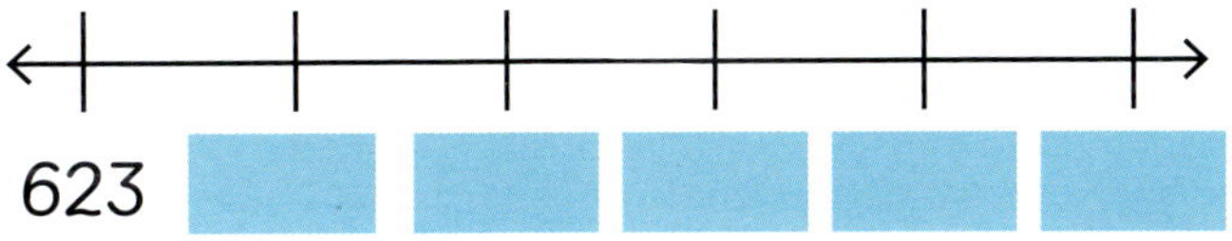

2 713 + 40 = ______

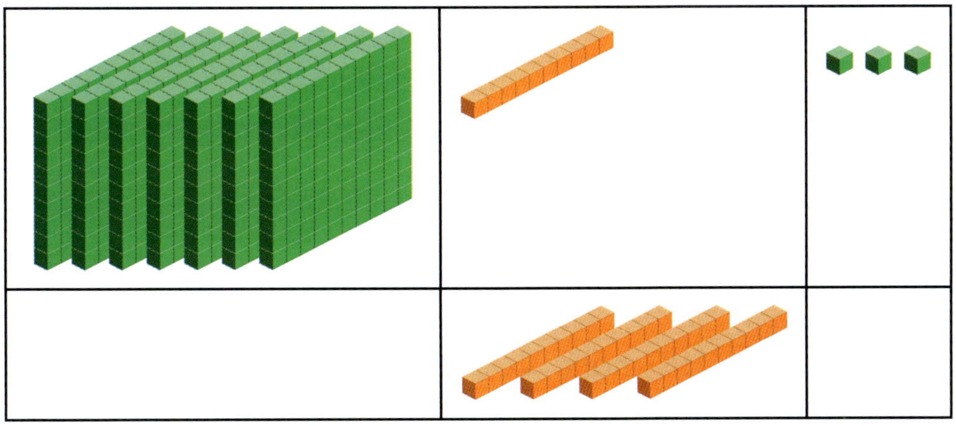

3 30 + 267 = ______

4 903 + 90 = ______

Complete Worksheet **3**: pages **29–30**

Adding Hundreds

Lesson 4

Explore

287 Summerville fans came to watch the game.
500 Wintertown fans also came to watch the game.
How many fans watched the game in total?

Master

287 + 500 =

Method 1

Count on in hundreds from 287.

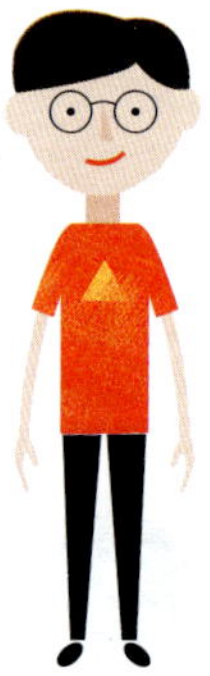

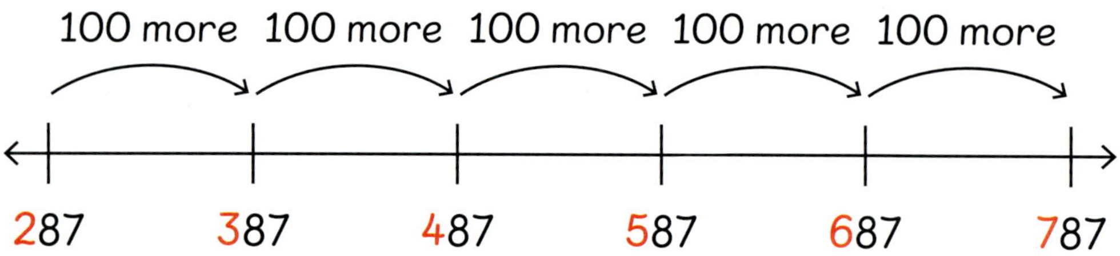

287 + 500 = 787

Method 2 Add the hundreds.

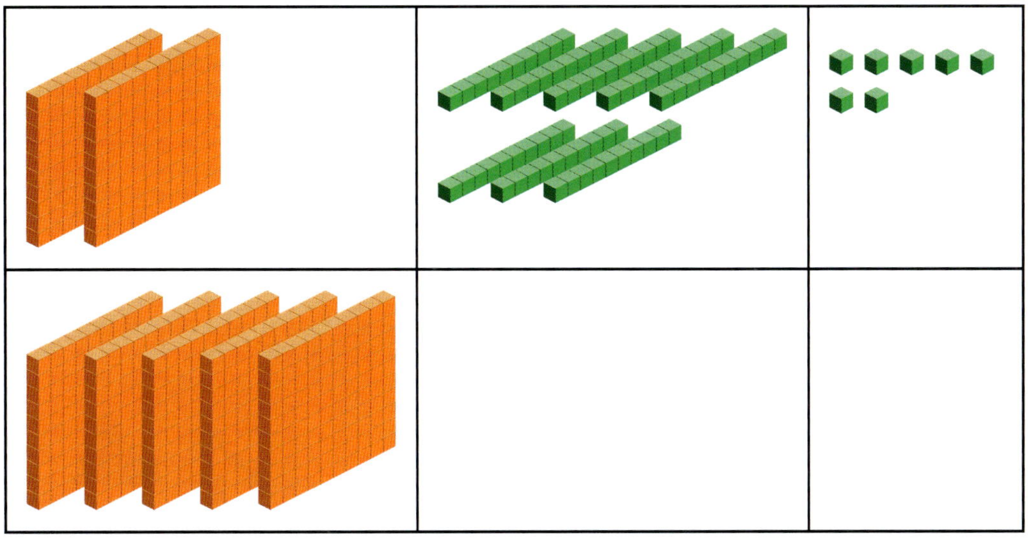

	h	t	o
	2	8	7
+	5	0	0
	7	8	7

287 + 500 = 787

787 fans watched the game in total.

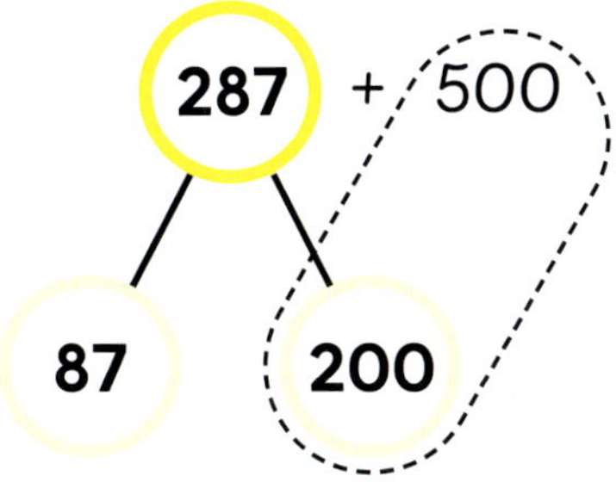

200 + 500 = 700

87 + 700 = 787

Guided Practice

Add.

1 146 + 700 = ☐

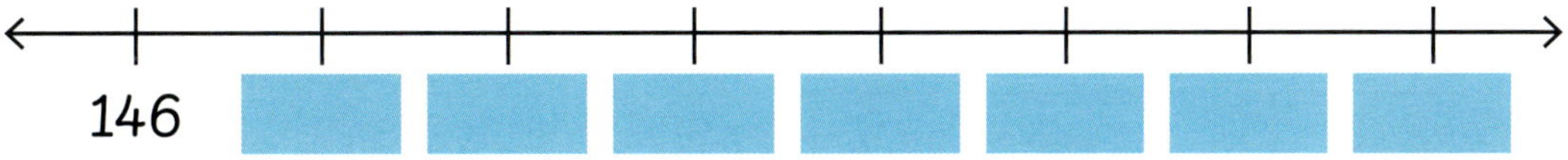

2 351 + 400 = ☐

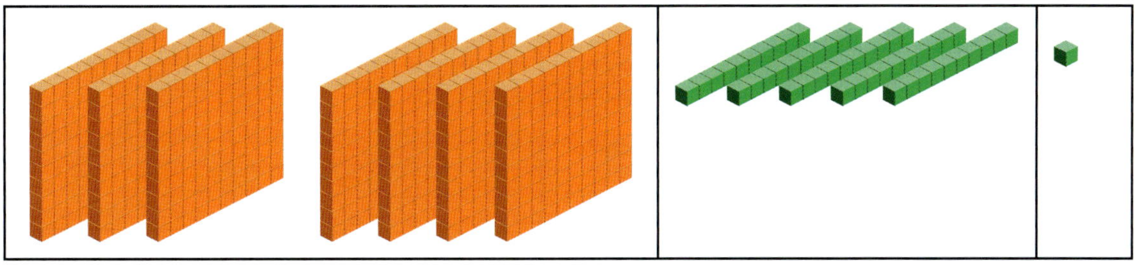

3 409 + 300 = ☐

4 600 + 273 = ☐

Complete Worksheet **4**: pages **31–32**

Simple Adding

Lesson
5

Explore

413 people visited the art gallery in the morning.
582 people visited the art gallery in the afternoon.
How many people visited the art gallery altogether?

Master

413 + 582 = ☐

Step 1 Add the ones.
3 ones + 2 ones = 5 ones

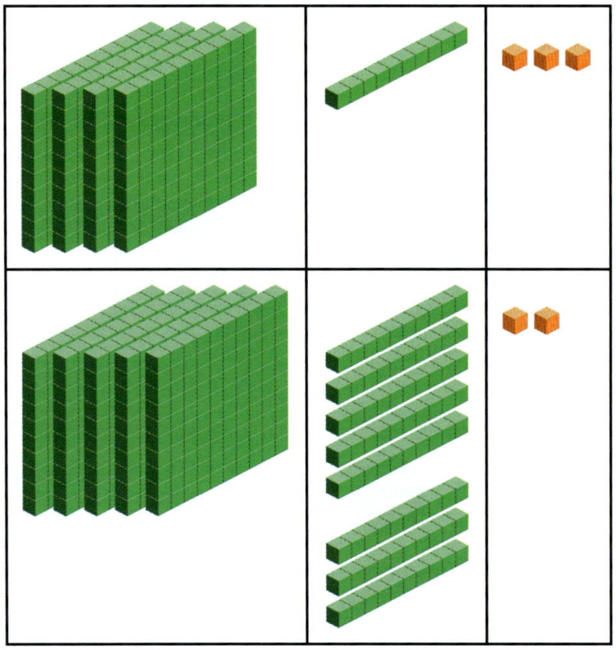

	h	t	o
	4	1	3
+	5	8	2
			5

Step 2 Add the tens.

1 ten + 8 tens = 9 tens

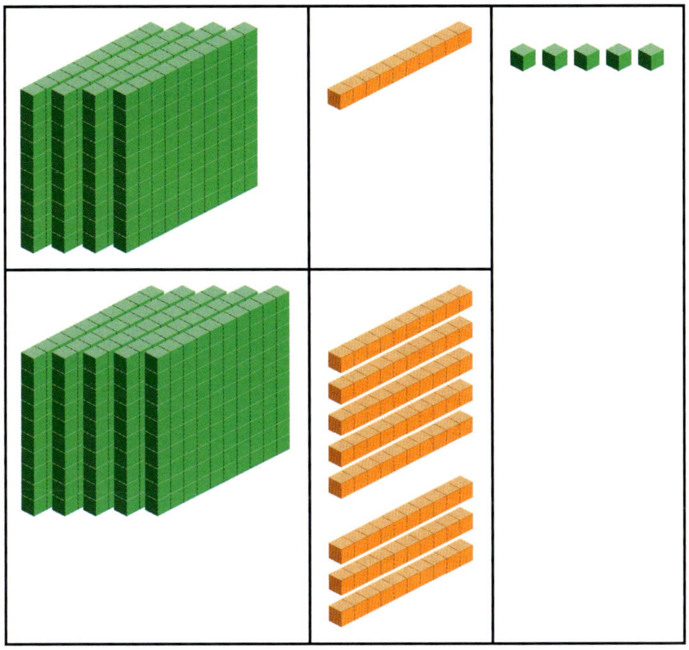

	h	t	o
	4	1	3
+	5	8	2
		9	5

Step 3 Add the hundreds.

4 hundreds + 5 hundreds = 9 hundreds

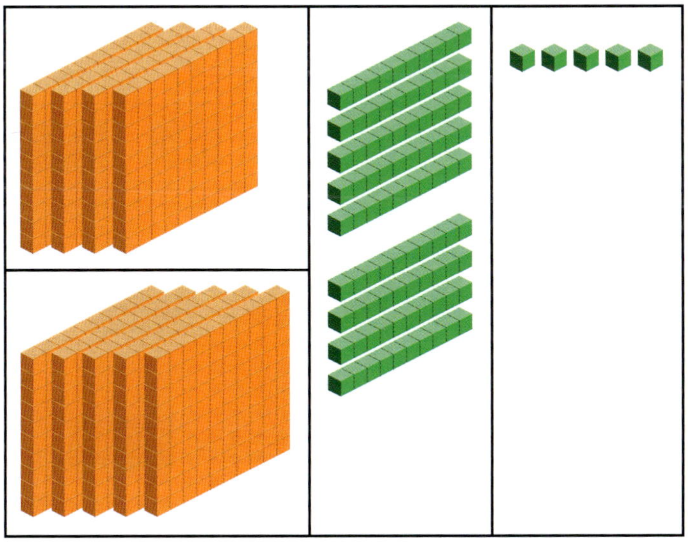

	h	t	o
	4	1	3
+	5	8	2
	9	9	5

413 + 582 = 995

995 people visited the art gallery altogether.

Guided Practice

Add.

1

(a) 134 + 3 = ☐

(b) 134 + 30 = ☐

(c) 134 + 300 = ☐

(d) 134 + 333 = ☐

2

(a) 213 + 4 = ☐

(b) 213 + 40 = ☐

(c) 213 + 400 = ☐

(d) 213 + 444 = ☐

3

(a)

	h	t	o
	4	3	6
+		2	3
	☐	☐	☐

(b)

	h	t	o
	3	1	4
+	4	5	3
	☐	☐	☐

(c) 215 + 42 = ☐

(d) 334 + 14 = ☐

(e) 412 + 267 = ☐

(f) 346 + 351 = ☐

Complete Worksheet **5**: pages **33–34**

Adding with Renaming

Lesson
6

Explore

7 people joined a group of 345 people at a fun run.
How many people were at the fun run in total?

Master

1 7 + 345 =

Method 1

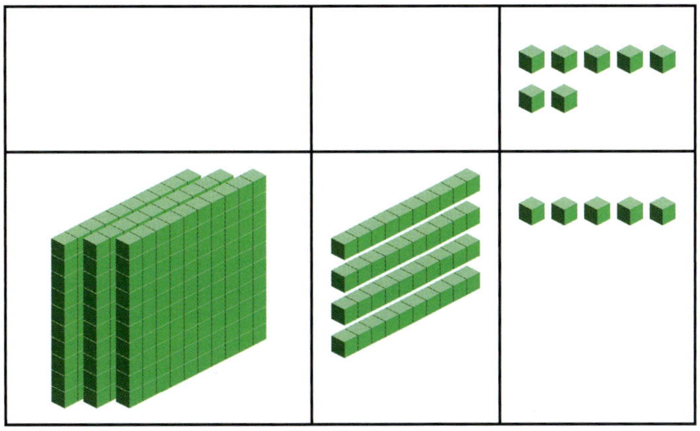

Step 1 Add the ones.

7 + 5 = 12

	h	t	o
			7
+	3	4	5
		1	2

Step 2 Add the tens.

0 + 40 = 40

	h	t	o
			7
+	3	4	5
		1	2
+		4	0

Step 3 Add the hundreds.

0 + 300 = 300

Step 4 Add 12, 40 and 300.

300 + 40 + 12 = 352

	h	t	o
			7
+	3	4	5
		1	2
		4	0
+	3	0	0
	3	5	2

7 + 345 = 352

Method 2

Step 1 Add the ones.

7 ones + 5 ones = 12 ones

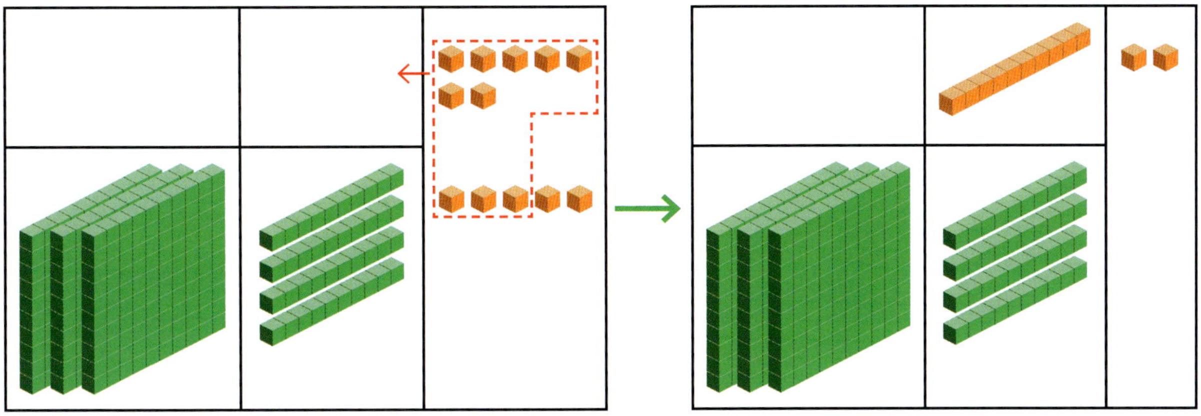

Rename the ones.

12 ones = 1 ten + 2 ones

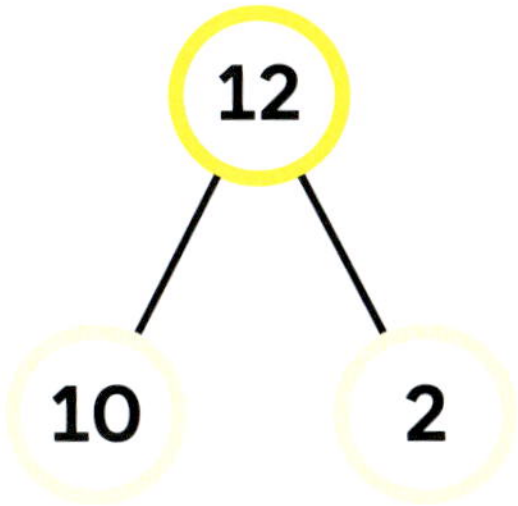

	h	t	o
			7
+	3	[1] 4	5
			2

Step 2 Add the tens.

1 ten + 4 tens = 5 tens

Add the hundreds.

0 hundreds + 3 hundreds = 3 hundreds

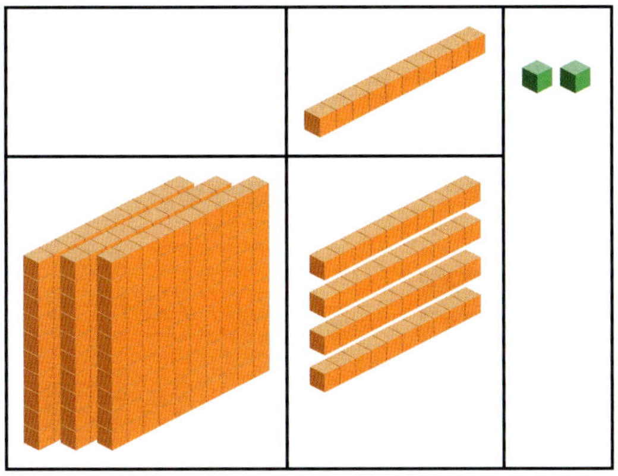

	h	t	o
			7
+	3	[1] 4	5
	3	5	2

7 + 345 = 352

Method 3

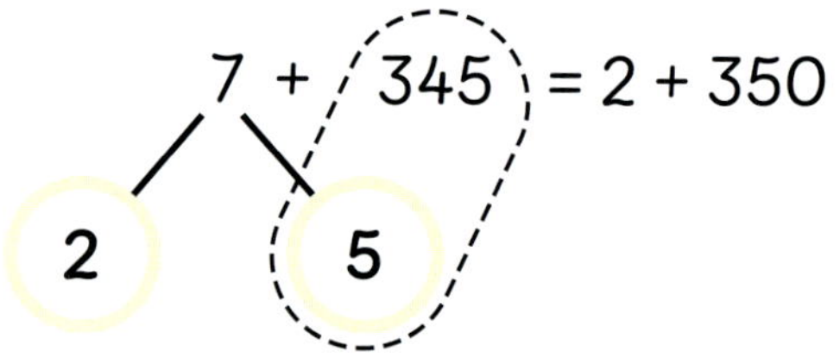

7 + 345 = 352

There were 352 people at the fun run in total.

2 53 + 47 = ▭

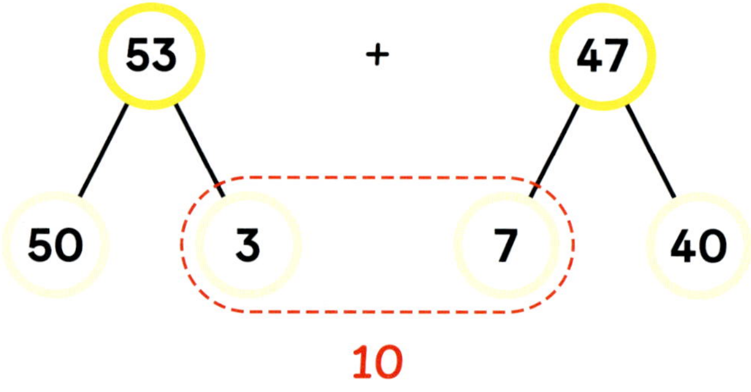

53 + 47 = 100

Guided Practice

Add.

1 (a) 5 + 7 = ▭ (b) 8 + 6 = ▭

(c) 4 + 9 = ▭ (d) 3 + 8 = ▭

2 (a) 35 + 7 = ▭ (b) 68 + 6 = ▭

(c) 54 + 9 = ▭ (d) 83 + 8 = ▭

3 (a) 8 + 457 = ▭ (b) 238 + 5 = ▭ (c) 729 + 6 = ▭

(d) 18 + 32 = ▭ (e) 63 + 37 = ▭ (f) 24 + 56 = ▭

Complete Worksheet **6**: pages **35–36**

Adding with Renaming

Lesson
7

Explore

Lulu, Sam and Hannah are playing a game. Their scores are shown below.

Lulu

Sam

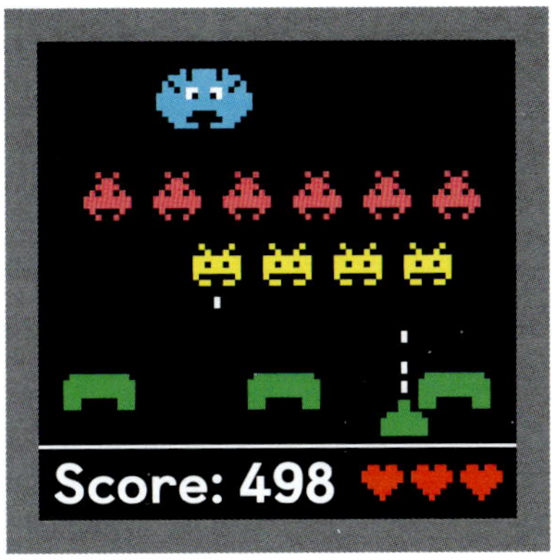

Hannah

They each played one more turn, then added their scores together.

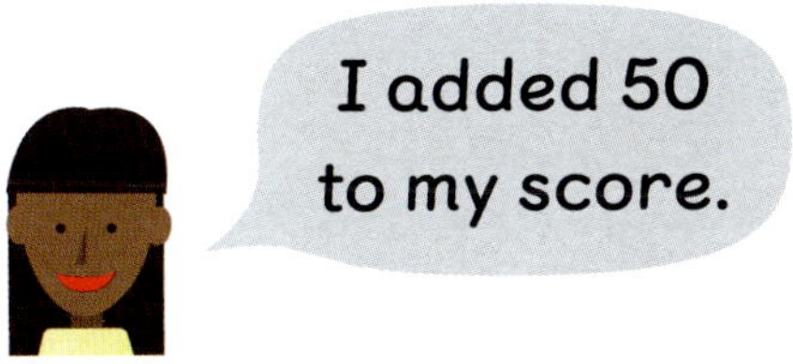

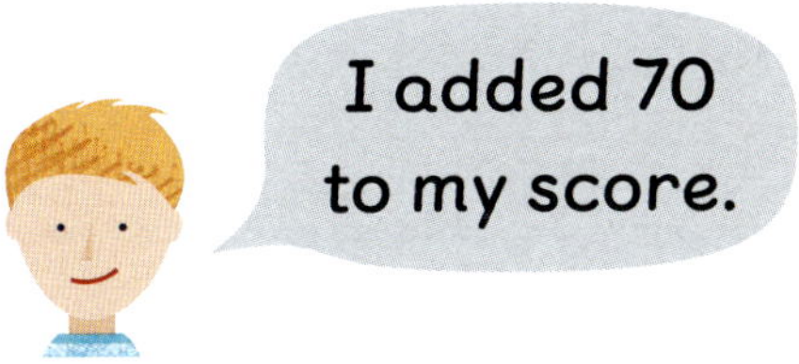

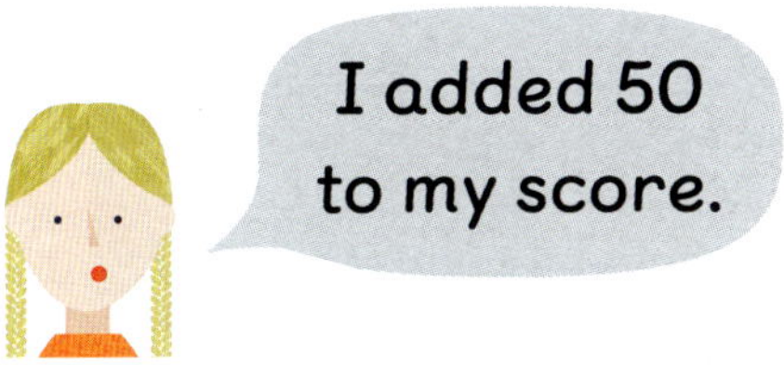

Who finished with the greatest score?

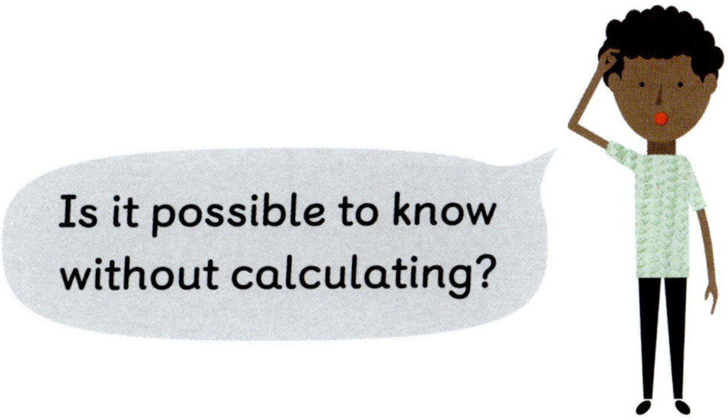

Master

1 593 + 50 = ______

Step 1 Add the ones.

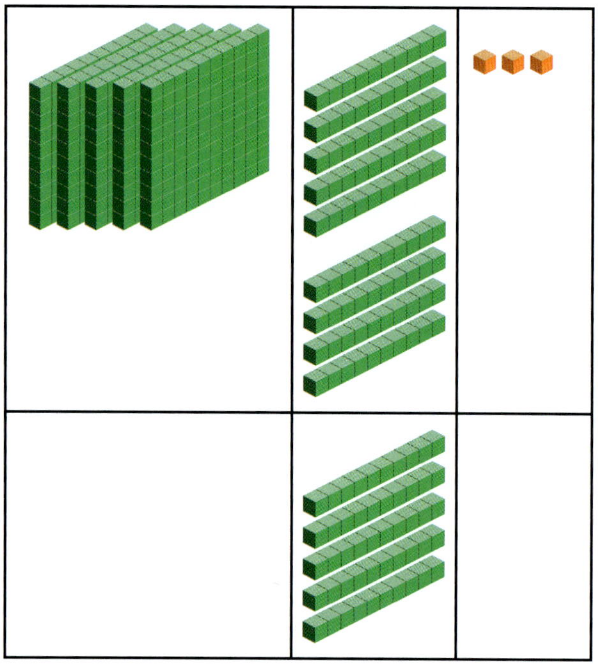

	h	t	o
	5	9	3
+		5	0
			3

Step 2 Add the tens.
9 tens + 5 tens = 14 tens

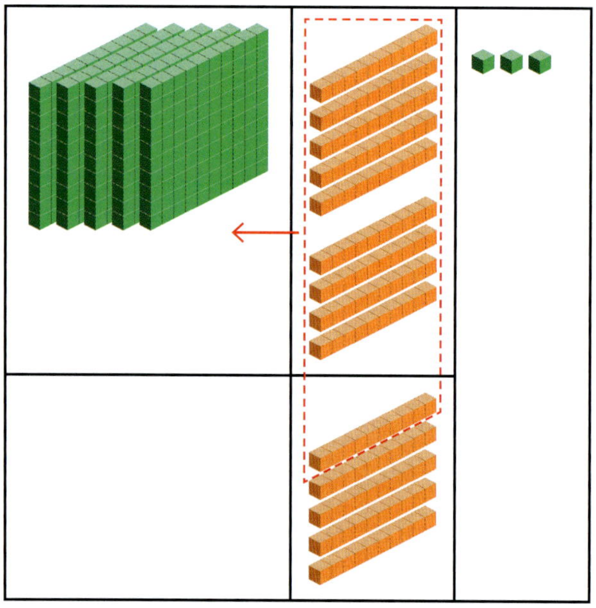

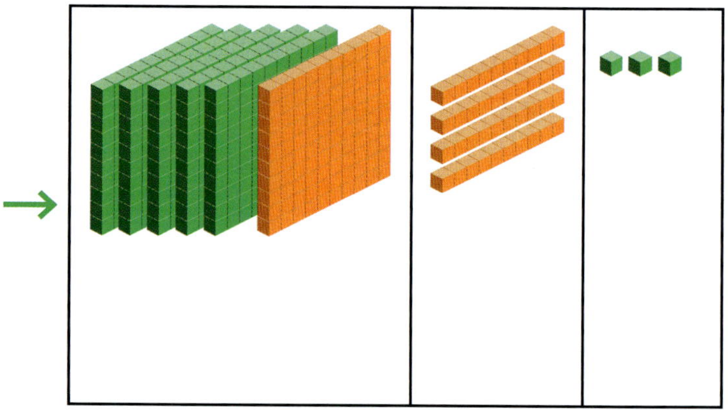

	h	t	o
	5	9	3
+		5	0
			3
	1	4	0

Rename the tens.
14 tens = 1 hundred + 4 tens

Step 3 Add the hundreds.

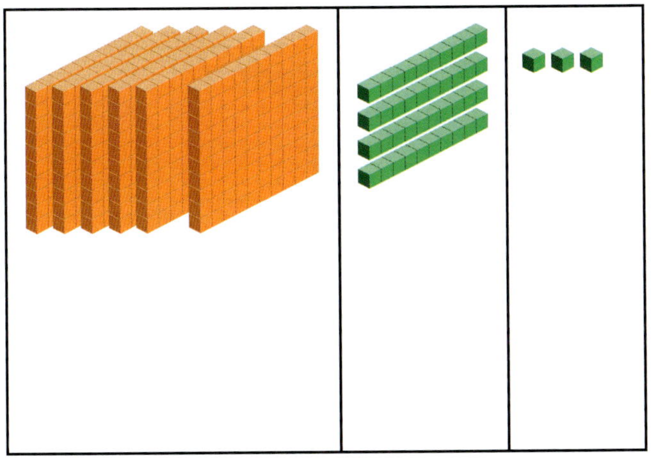

	h	t	o
	5	9	3
+		5	0
			3
	1	4	0
+	5	0	0
	6	4	3

Step 4 Add 3, 140 and 500

500 + 140 + 3 = 643

593 + 50 = 643

2 695 + 70 =

Step 1 Add the ones.

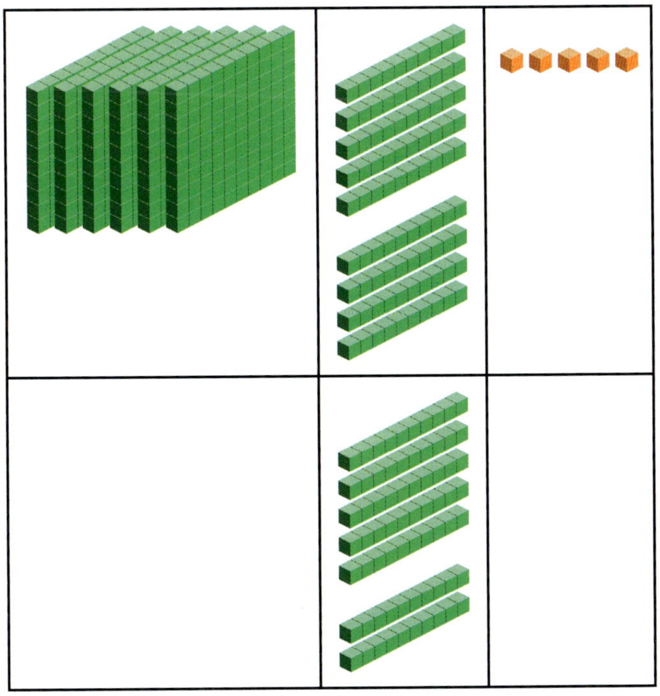

	h	t	o
	6	9	5
+		7	0
			5

Step 2 Add the tens.

9 tens + 7 tens = 16 tens

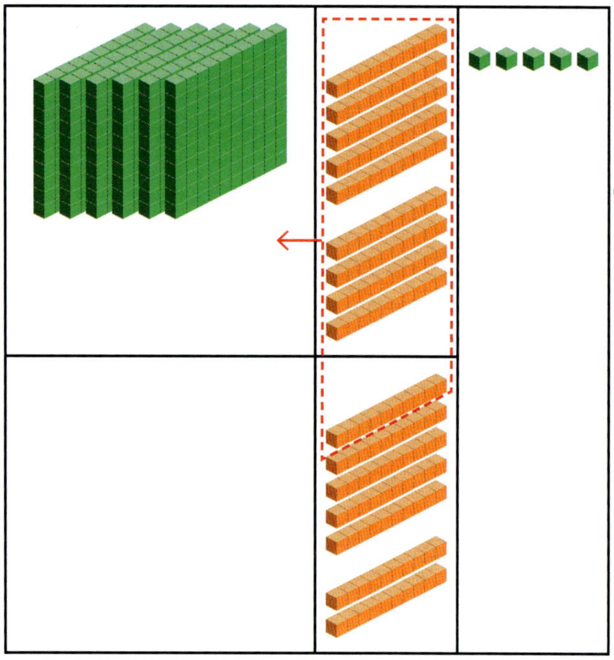

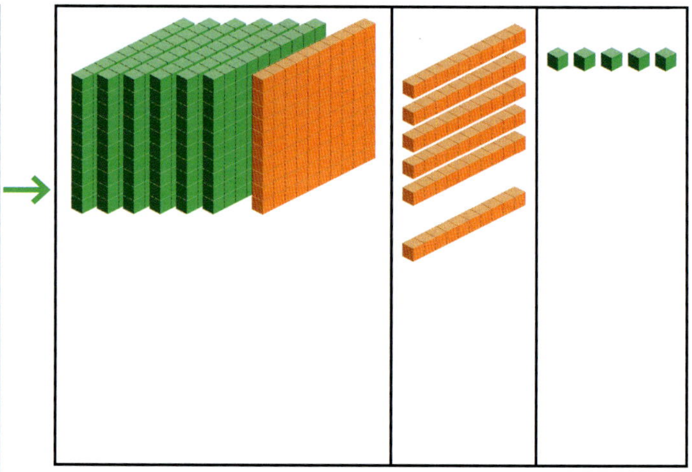

h	t	o
[1] 6	9	5
+	7	0
	6	5

Rename the tens.

16 tens = 1 hundred + 6 tens

Step 3 Add the hundreds.

1 hundred + 6 hundreds = 7 hundreds

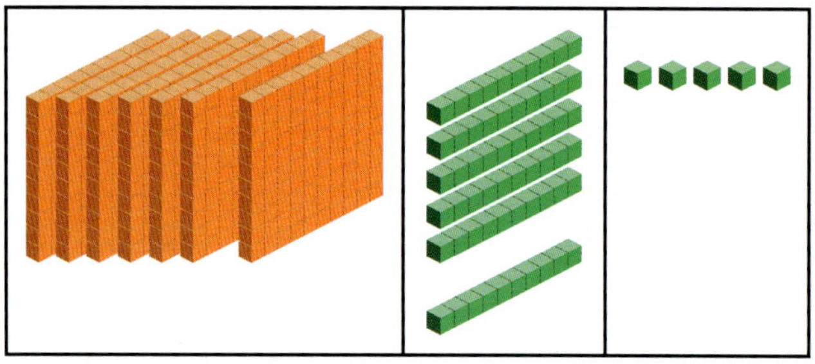

h	t	o
[1] 6	9	5
+	7	0
7	6	5

695 + 70 = 765

3 498 + 50 =

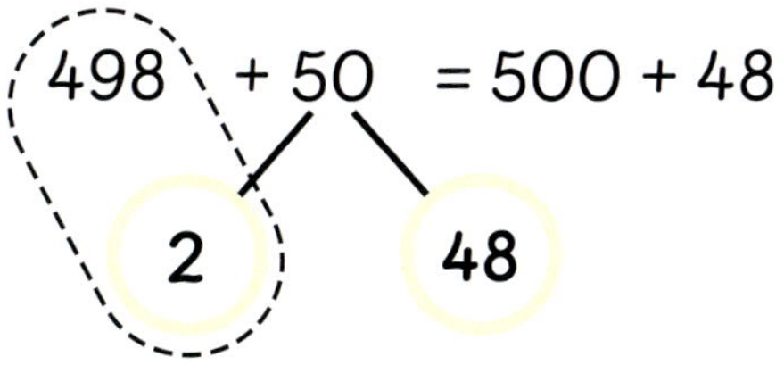

498 + 50 = 548

593 + 50 = 643	Lulu has 643 points.
695 + 70 = 765	Sam has 765 points.
498 + 50 = 548	Hannah has 548 points.

4

Lulu

Sam

Hannah

600 + 50 = 650

700 + 70 = 770

500 + 50 = 550

Lulu now has about 650 points.
Sam now has about 770 points.
Hannah now has about 550 points.

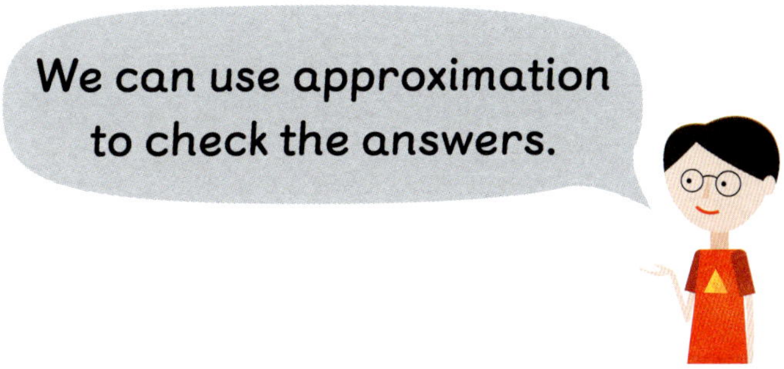

Sam finished with the greatest score.

Guided Practice

1 Roll a dice to fill in the first blank, then complete the equation.

(a) 693 + ____ 0 = ____

(b) 693 + ____ 0 = ____

(c) 693 + ____ 0 = ____

2 Estimate the total.

(a) 498 + 70

498 is about ____.

498 + 70 is about ____.

(b) 20 + 796

796 is about ____.

20 + 796 is about ____.

(c) 895 + 60

895 is about ____.

895 + 60 is about ____.

Complete Worksheet 7: pages 37–38

Adding with Renaming

Lesson 8

Explore

257 people borrowed books from the library on Monday.
315 people borrowed books on Tuesday.
How many people borrowed books over the two days?

Master

257 + 315 =

Step 1 Add the ones.
7 ones + 5 ones = 12 ones

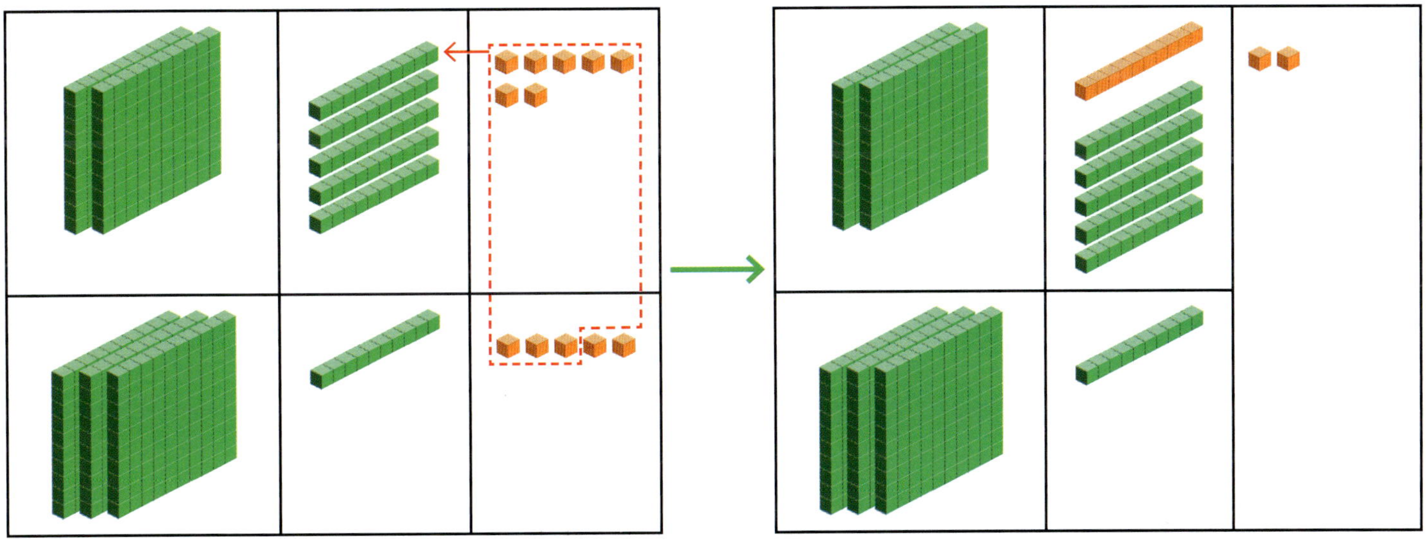

Rename the ones.
12 ones = 1 ten + 2 ones

	h	t	o
	2	$\overset{1}{5}$	7
+	3	1	5
			2

Step 2 Add the tens.

1 ten + 5 tens + 1 ten = 7 tens

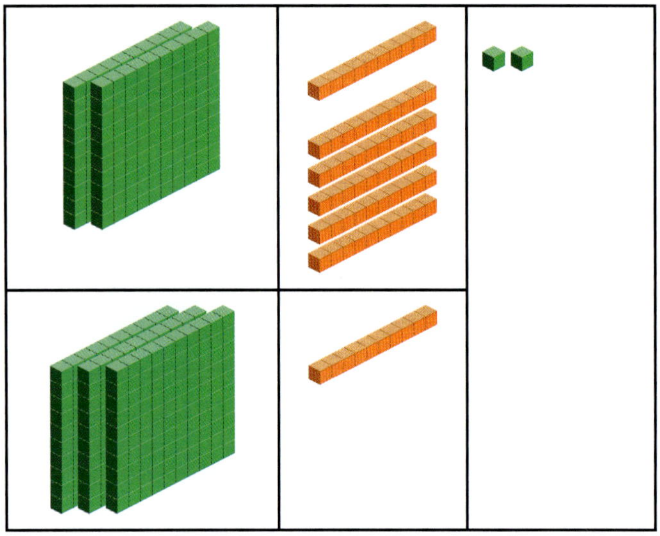

	h	t	o
		1	
	2	5	7
+	3	1	5
		7	2

Step 3 Add the hundreds.

2 hundreds + 3 hundreds = 5 hundreds

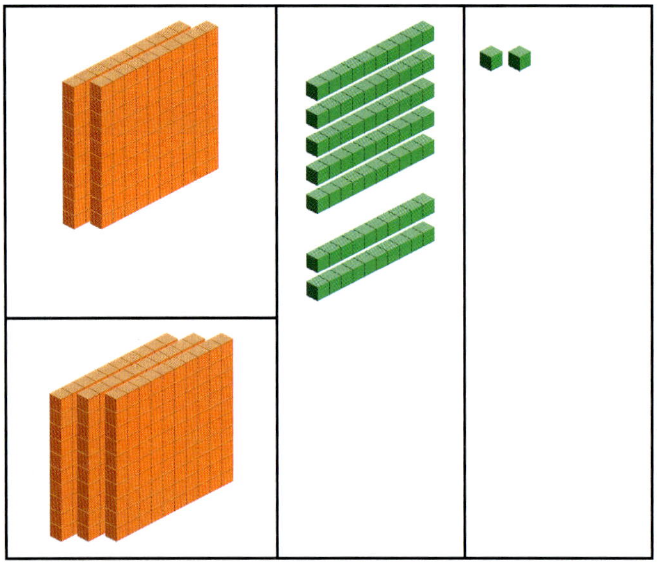

	h	t	o
		1	
	2	5	7
+	3	1	5
	5	7	2

257 + 315 = 572

572 people borrowed books over the two days.

Guided Practice

Add.

1

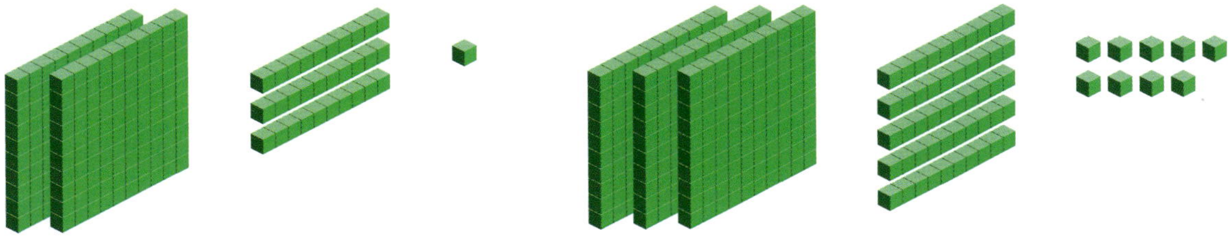

☐ + ☐ = ☐

2

(a) 534 + 127 = ☐

	h	t	o
	5	3	4
+	1	2	7
	☐	☐	☐

(b) 759 + 238 = ☐

	h	t	o
	7	5	9
+	2	3	8
	☐	☐	☐

(c) 375 + 216 = ☐

	h	t	o
	3	7	5
+	2	1	6
	☐	☐	☐

(d) 48 + 436 = ☐

	h	t	o
		4	8
+	4	3	6
	☐	☐	☐

3

(a) 52 + 138 = ☐

(b) 173 + 409 = ☐

Complete Worksheet **8**: pages **39–40**

Adding with Renaming

Lesson 9

Explore

482 people attended an ice hockey match on Friday night.
365 people attended an ice hockey match on Saturday night.
How many people in total attended an ice hockey match over the two nights?

Master

482 + 365 =

Step 1 Add the ones.
2 ones + 5 ones = 7 ones

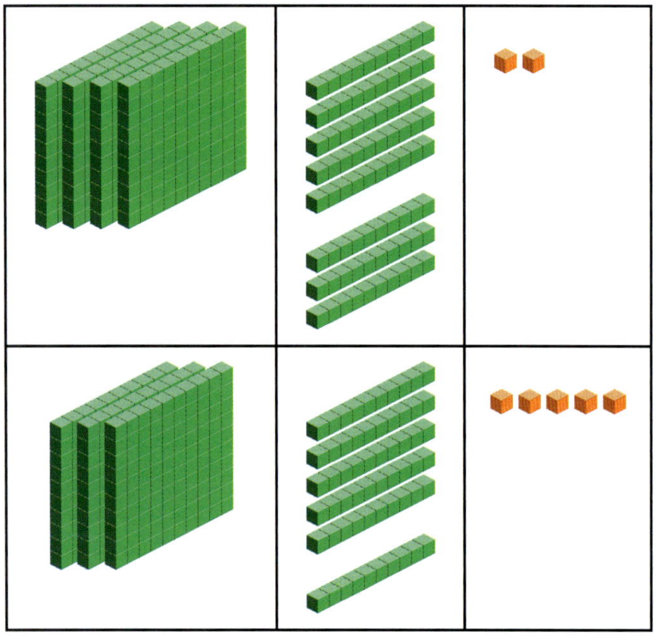

	h	t	o
	4	8	2
+	3	6	5
			7

Step 2 Add the tens.
8 tens + 6 tens = 14 tens

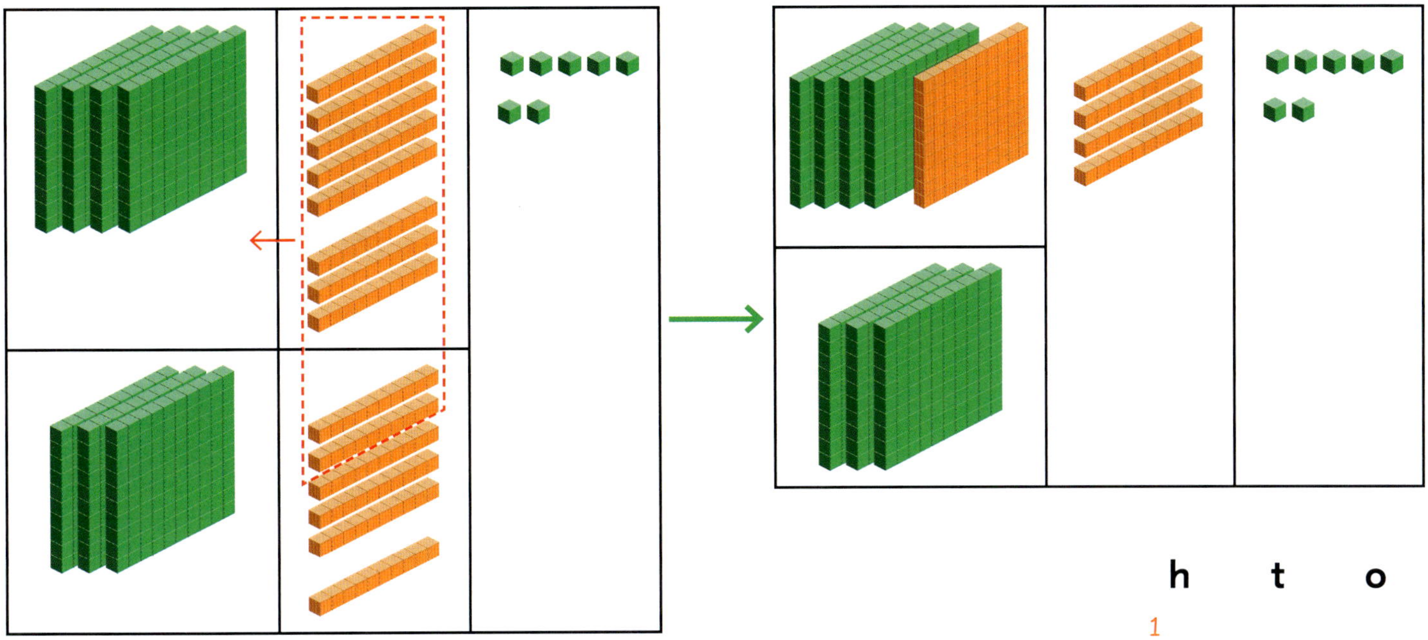

Rename the tens.
14 tens = 1 hundred + 4 tens

	h	t	o
	^{1}4	8	2
+	3	6	5
		4	7

Step 3 Add the hundreds.
1 hundred + 4 hundreds + 3 hundreds = 8 hundreds

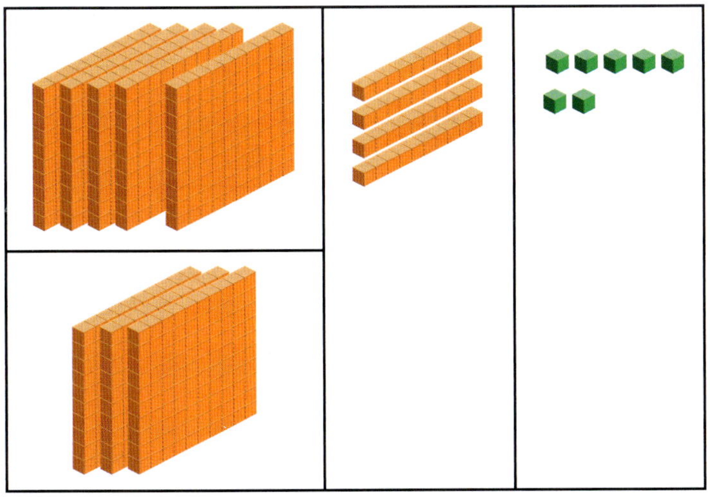

	h	t	o
	^{1}4	8	2
+	3	6	5
	8	4	7

482 + 365 = 847

847 people in total attended an ice hockey match over the two nights.

Guided Practice

Add.

1

+ =

2

(a) 324 + 392 =

	h	t	o
	3	2	4
+	3	9	2

(b) 572 + 361 =

	h	t	o
	5	7	2
+	3	6	1

(c) 476 + 143 =

	h	t	o
	4	7	6
+	1	4	3

(d) 41 + 267 =

	h	t	o
		4	1
+	2	6	7

3

(a) 85 + 564 =

(b) 193 + 421 =

Complete Worksheet **9**: pages **41–42**

Adding with Renaming

Lesson 10

Explore

The white lorry has to deliver 583 boxes.
The blue lorry has to deliver 249 boxes.
How many boxes do they have to deliver altogether?

Master

583 + 249 = ▭

Step 1 Add the ones.
3 ones + 9 ones = 12 ones

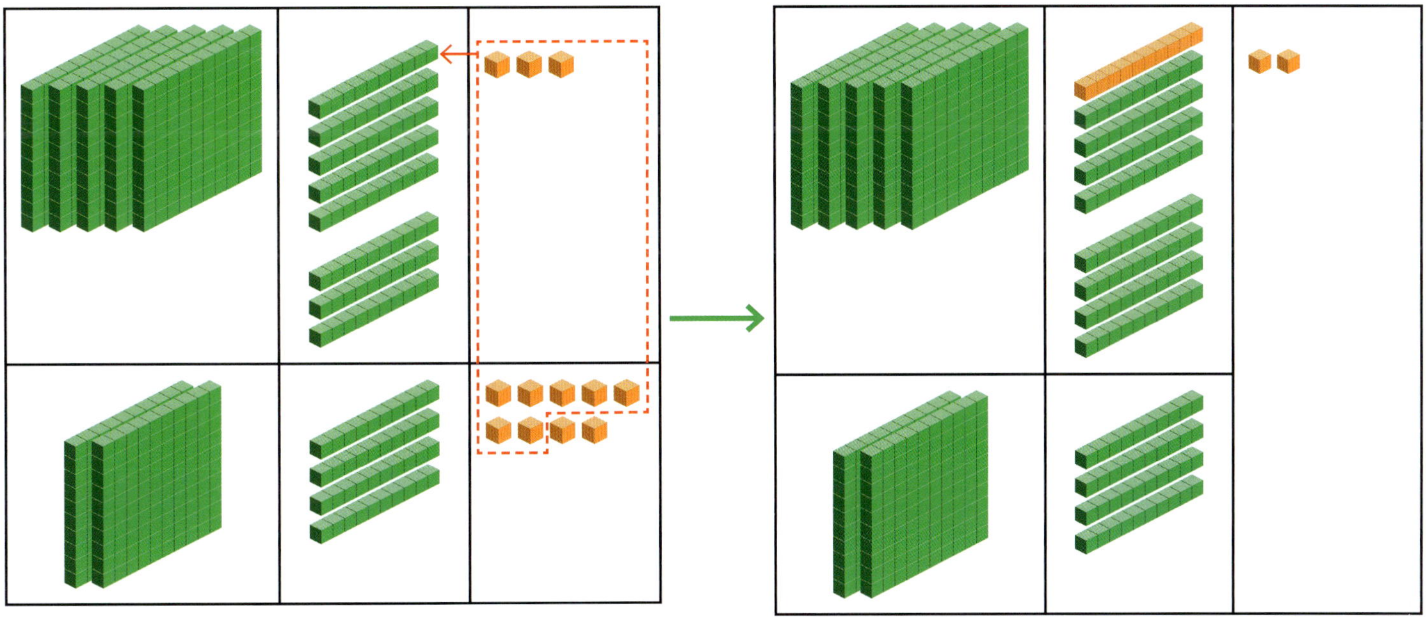

Rename the ones.
12 ones = 1 ten + 2 ones

	h	t	o
		1	
	5	8	3
+	2	4	9
			2

Step 2 Add the tens.

1 ten + 8 tens + 4 tens = 13 tens

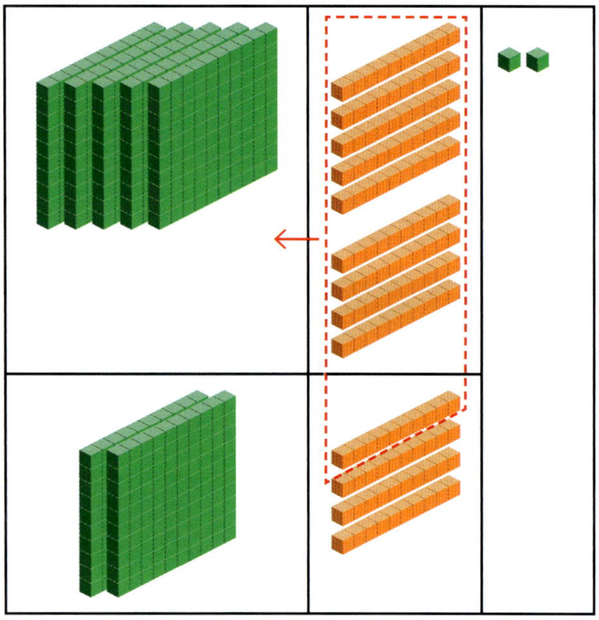

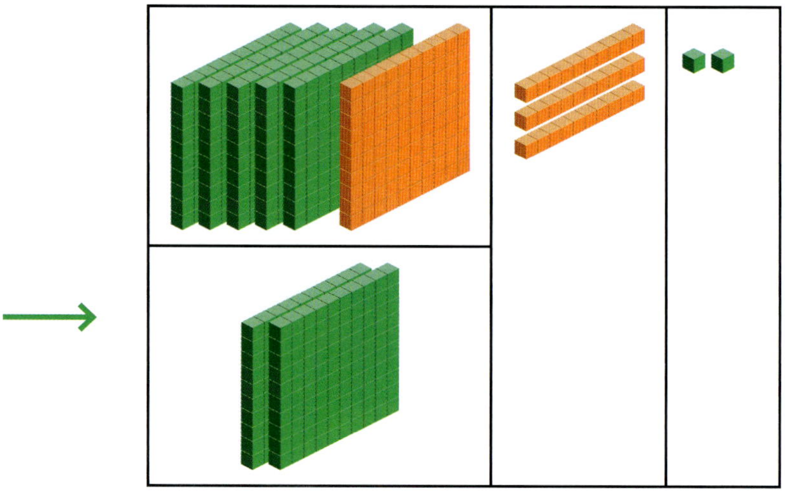

Rename the tens.

13 tens = 1 hundred + 3 tens

	h	t	o
	[1] 5	[1] 8	3
+	2	4	9
		3	2

Step 3 Add the hundreds.

1 hundred + 5 hundreds + 2 hundreds = 8 hundreds

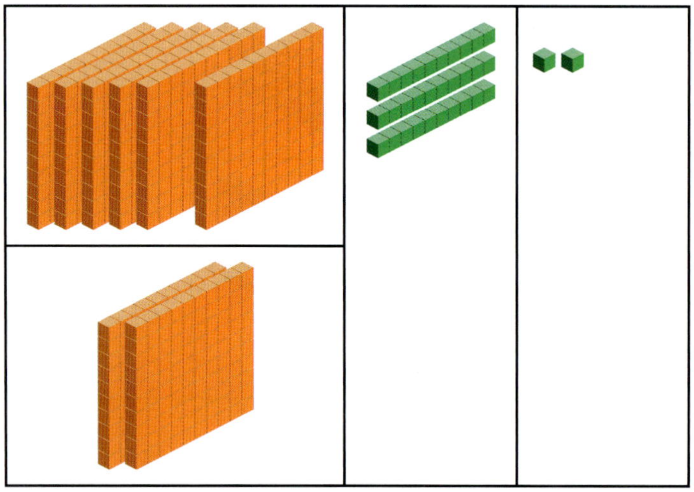

	h	t	o
	[1] 5	[1] 8	3
+	2	4	9
	8	3	2

583 + 249 = 832

The lorries have to deliver 832 boxes altogether.

Activity Time

Partner work

You will need:

1. Roll three times to get three digits.
2. Put the digits in any order to form an equation:

389 + ☐ ☐ ☐ = ☐

3. The aim is to make a total as close as possible to 700.
4. Repeat the activity, taking turns.

Guided Practice

Add.

1 (a) 248 + 174 = ☐

	h	t	o
	2	4	8
+	1	7	4
	☐	☐	☐

(b) 367 + 359 = ☐

	h	t	o
	3	6	7
+	3	5	9
	☐	☐	☐

2 (a) 486 + 368 = ☐

	h	t	o
	4	8	6
+	3	6	8
	☐	☐	☐

(b) 74 + 369 = ☐

	h	t	o
		7	4
+	3	6	9
	☐	☐	☐

Complete Worksheet **10**: pages **43–44**

Subtracting Ones

Lesson 11

Explore

Ravi collected 587 football cards.
He gave 5 football cards to Amira.
How many football cards does Ravi have left?

Master

$587 - 5 = \square$

Method 1 Count back from 587.

$587 - 5 = 582$

Method 2 Subtract the ones.

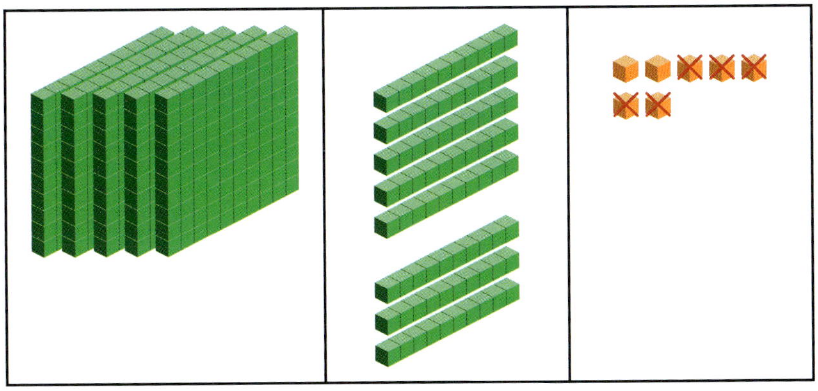

	h	t	o
	5	8	7
–			5
	5	8	2

587 – 5 = 582

Ravi has 582 football cards left.

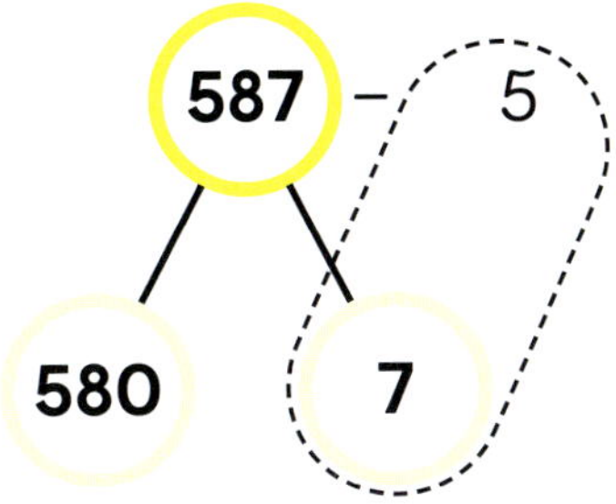

7 – 5 = 2

580 + 2 = 582

Guided Practice

Subtract.

1

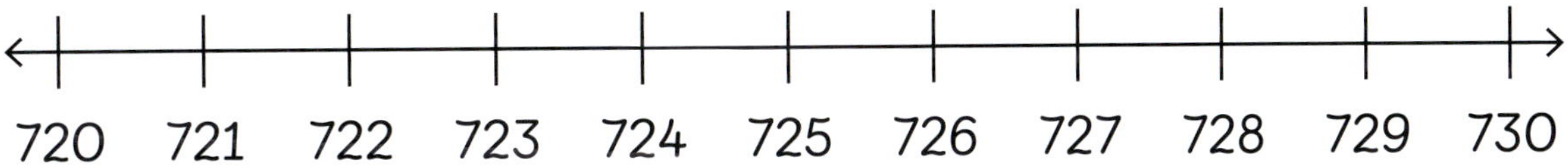

(a) 726 – 3 =

(b) 729 – 7 =

2 (a) 507 – 5 =

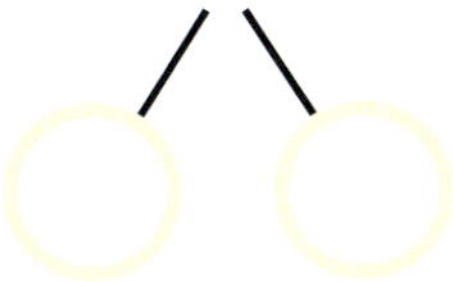

(b) 608 – 8 =

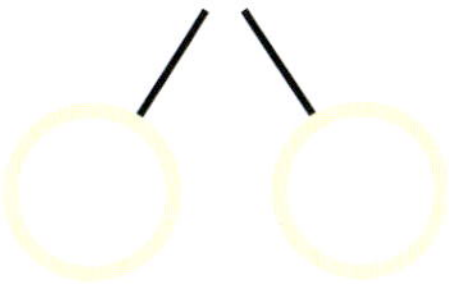

3 (a) 76 – 4 =

(b) 176 – 4 =

(c) 87 – 3 =

(d) 387 – 3 =

Complete Worksheet 11: pages 45–46

Subtracting Tens

Lesson 12

Explore

Holly has 279 football stickers in a box.
She pastes 50 stickers in a sticker album.
How many stickers does she have left in the box?

Master

279 – 50 =

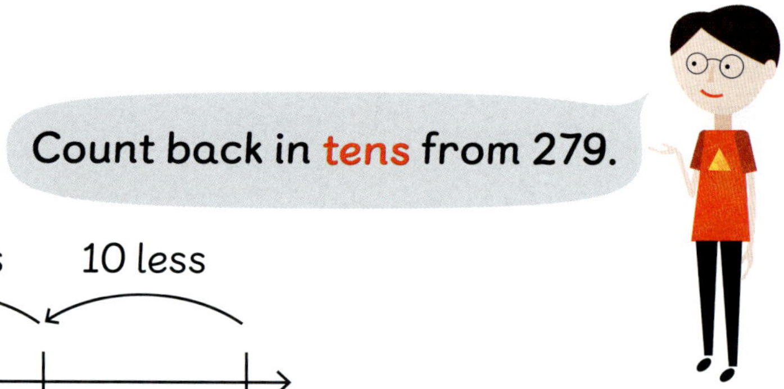

Method 1

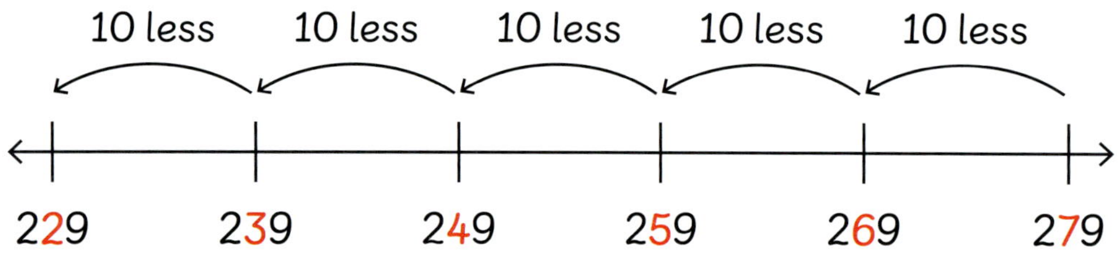

279 – 50 = 229

Method 2 Subtract the tens.

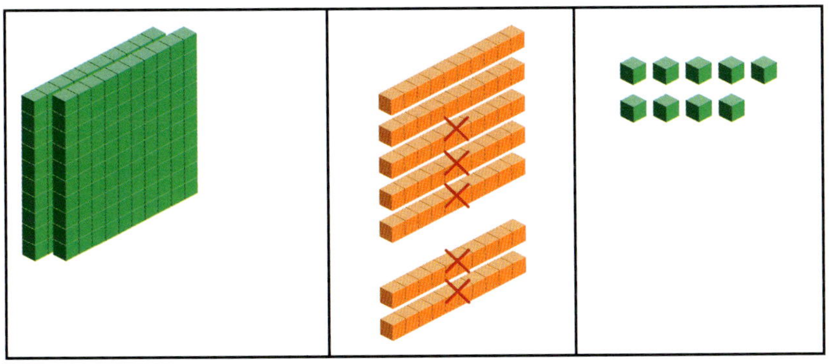

h	t	o
2	7	9
−	5	0
2	2	9

279 − 50 = 229

Holly has 229 stickers left.

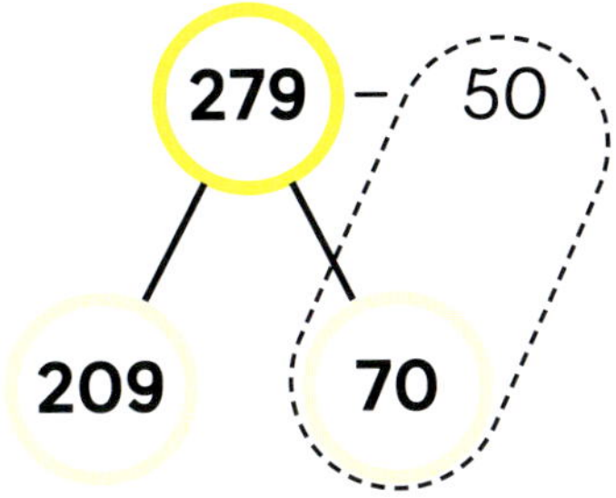

70 − 50 = 20
209 + 20 = 229

Guided Practice

Subtract.

1 856 − 30 = ____

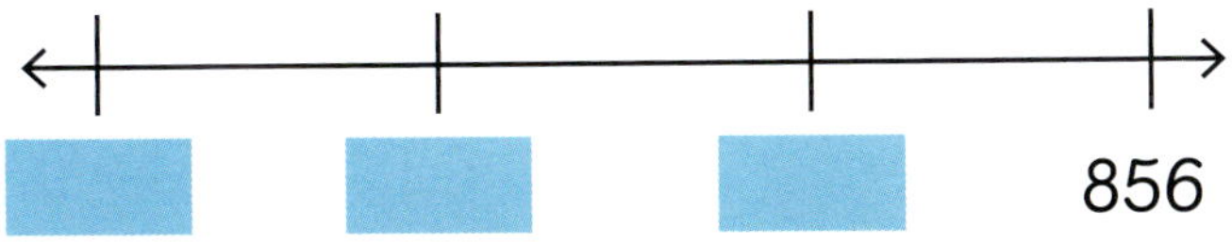

2 567 − 40 = ____

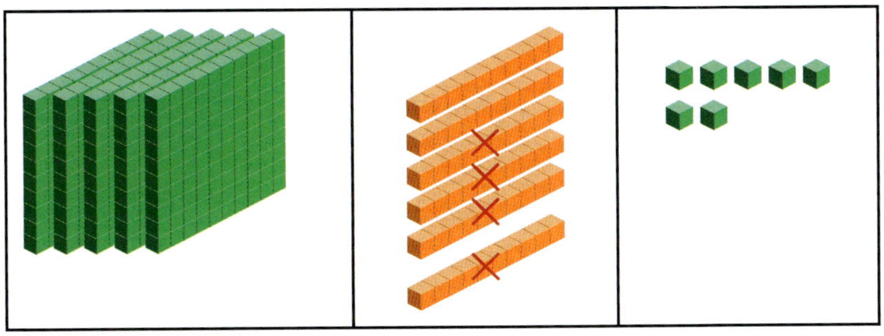

3 (a) 798 − 6 = ____ (b) 798 − 60 = ____

Complete Worksheet **12**: pages **47–48**

Subtracting Hundreds

Lesson 13

Explore

796 people were at the airport.
600 people got on their flight.
How many people were left at the airport?

Master

796 − 600 =

Method 1

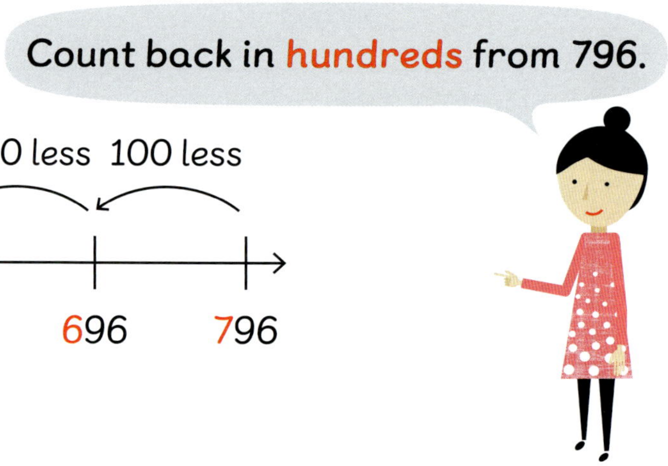

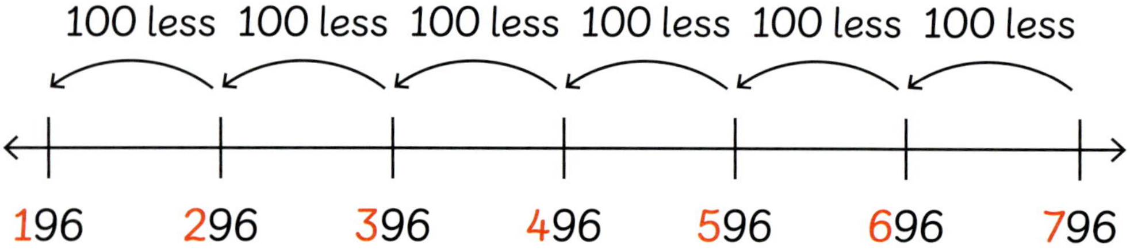

796 − 600 = 196

Method 2 Subtract the hundreds.

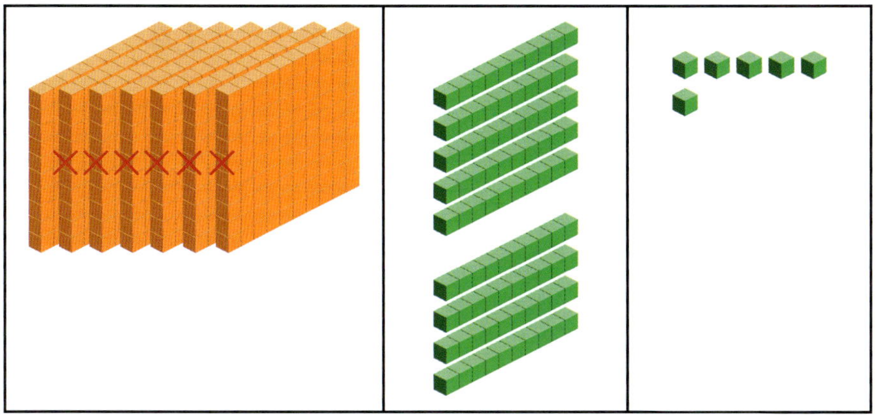

	h	t	o
	7	9	6
−	6	0	0
	1	9	6

796 − 600 = 196

There were 196 people left at the airport.

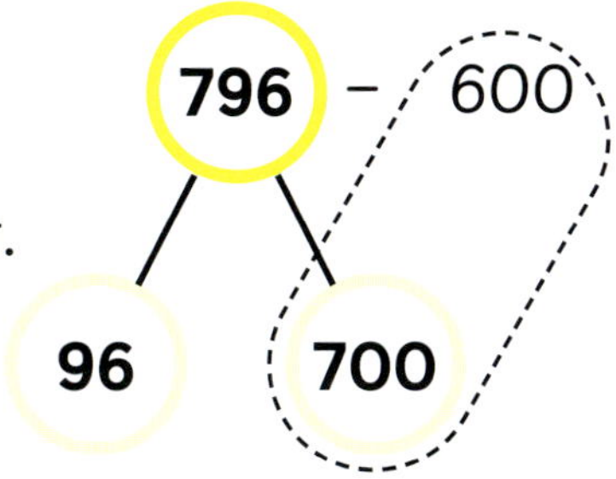

700 − 600 = 100
96 + 100 = 196

Guided Practice

Subtract.

1 743 − 400 =

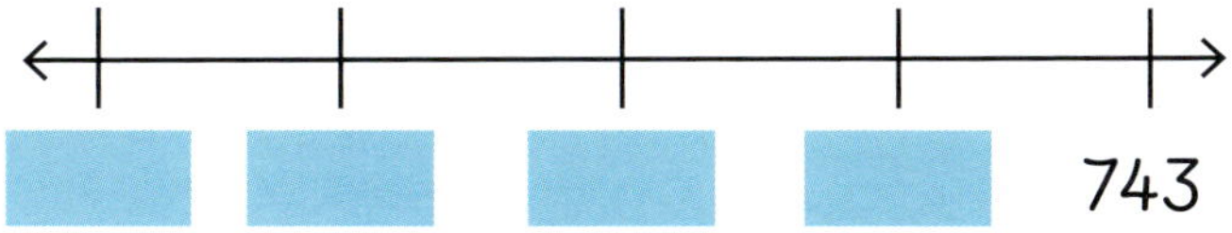

2 521 − 300 =

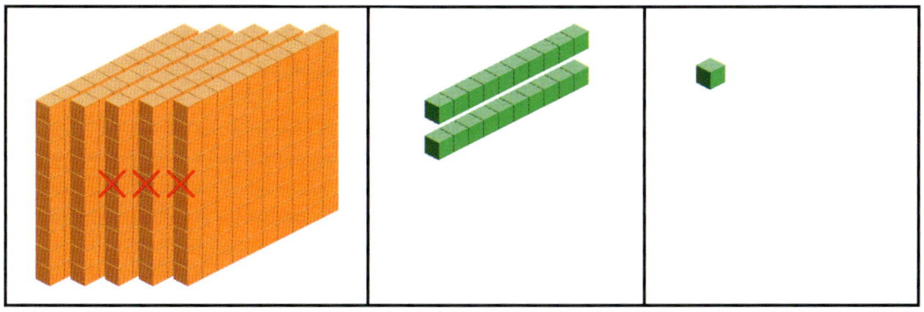

3 (a) 679 − 5 = (b) 679 − 50 = (c) 679 − 500 =

Complete Worksheet **13**: pages **49–50**

Simple Subtracting

Lesson 14

Explore

A restaurant has 748 tomatoes. It uses 425 tomatoes during one week to make its famous ketchup. How many tomatoes are left?

Master

748 − 425 =

Step 1 Subtract the ones.
8 ones − 5 ones = 3 ones

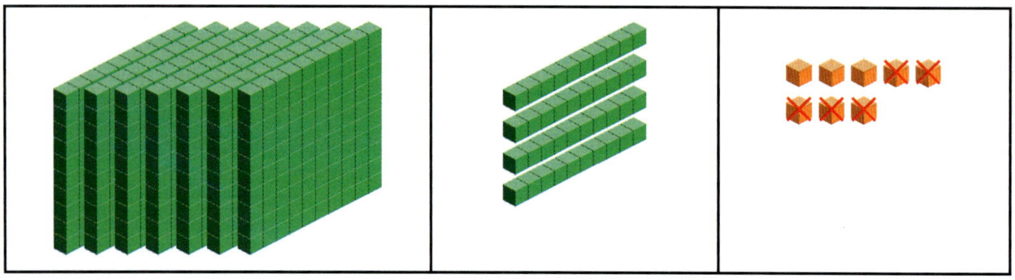

	h	t	o
	7	4	8
−	4	2	5
			3

Step 2 Subtract the tens.

4 tens − 2 tens = 2 tens

	h	t	o
	7	4	8
−	4	2	5
		2	3

Step 3 Subtract the hundreds.

7 hundreds − 4 hundreds = 3 hundreds

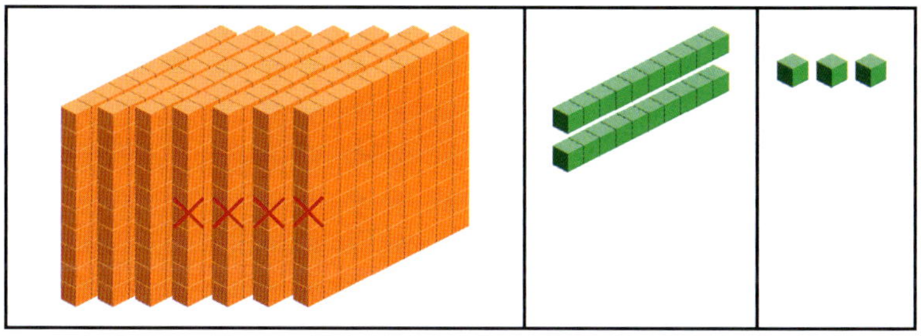

	h	t	o
	7	4	8
−	4	2	5
	3	2	3

748 − 425 = 323

323 tomatoes are left.

Guided Practice

Subtract.

1

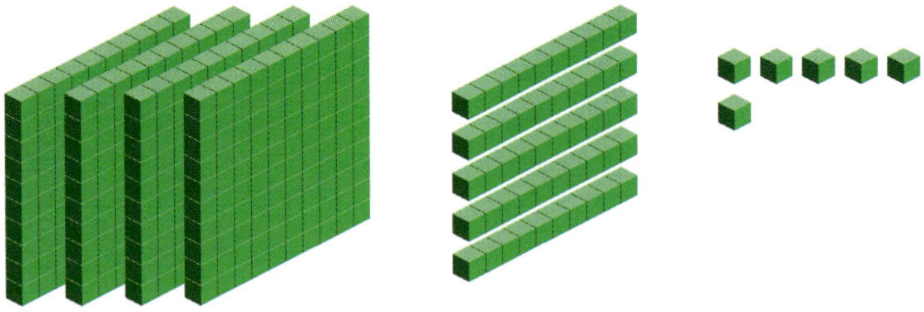

456 − 243 = ☐

2 564 − 231 = ☐

	h	t	o
	5	6	4
−	2	3	1
	☐	☐	☐

3 759 − 305 = ☐

	h	t	o
	7	5	9
−	3	0	5
	☐	☐	☐

4 (a) 834 − 431 = ☐ (b) 608 - 402 = ☐

5 (a) 354 − 21 = ☐ (b) 678 - 45 = ☐

6 (a) 452 − ☐ = 52 (b) 375 − ☐ = 175

Complete Worksheet **14**: pages **51–52**

Subtracting with Renaming

Lesson 15

Explore

What does Sam need to do to find the answer?

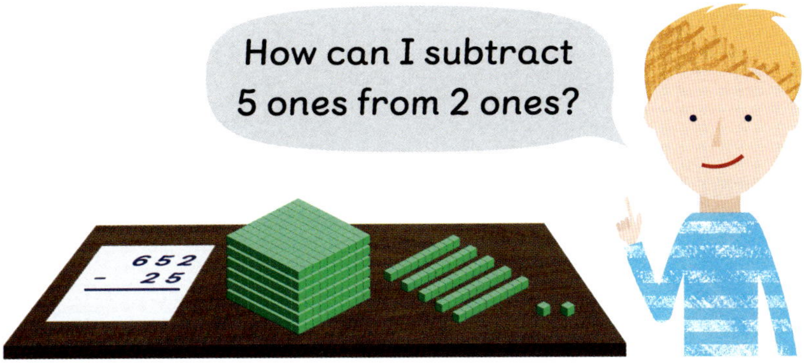

Master

652 – 25 = ▭

Step 1 Rename 1 ten as 10 ones.
Subtract the ones.

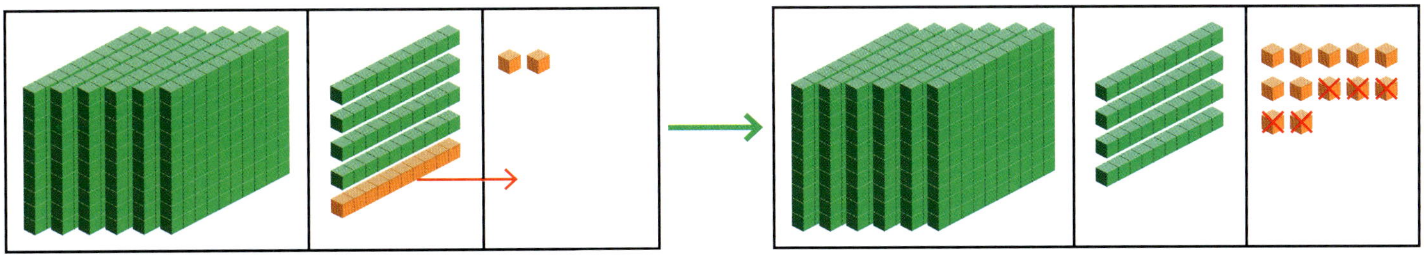

12 ones – 5 ones = 7 ones

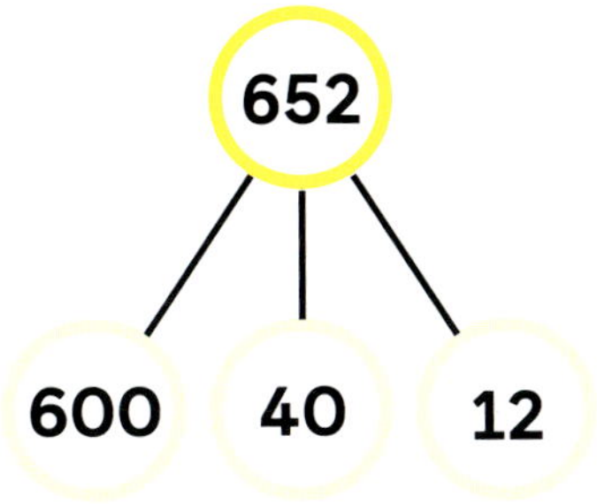

	h	t	o
		4	12
	6	~~5~~	~~2~~
–		2	5
			7

Step 2 Subtract the tens.

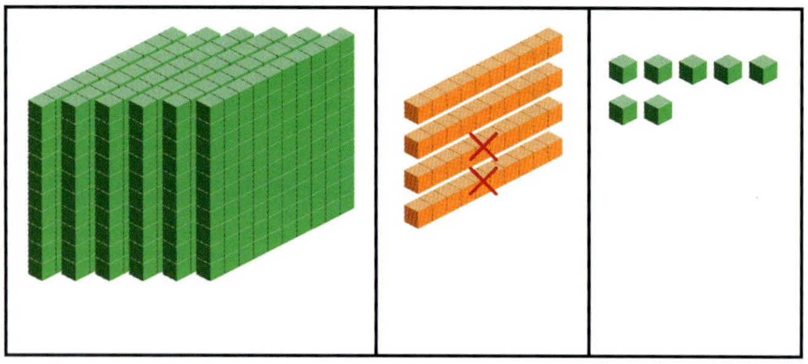

	h	t	o
		4	12
	6	~~5~~	~~2~~
–		2	5
		2	7

4 tens – 2 tens = 2 tens

Step 3 Subtract the hundreds.

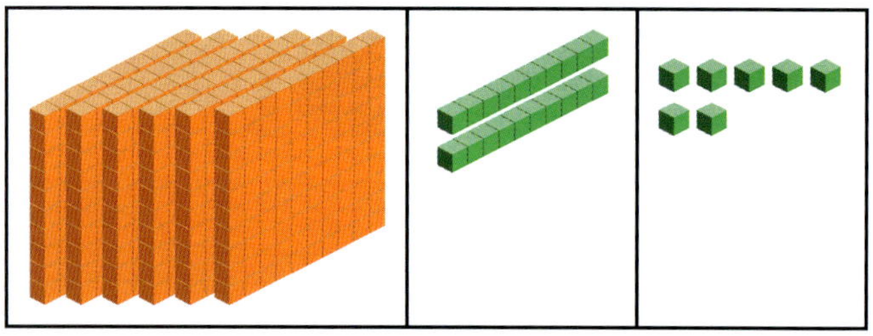

	h	t	o
		4	12
	6	~~5~~	~~2~~
–		2	5
	6	2	7

6 hundreds – 0 hundreds = 6 hundreds

652 – 25 = 627

Guided Practice

Subtract.

1 578 – 39 = ☐

2 743 – 26 = ☐

3 876 – 328 = ☐

4 645 – 429 = ☐

Complete Worksheet **15**: pages **53–54**

Subtracting with Renaming

Lesson 16

Explore

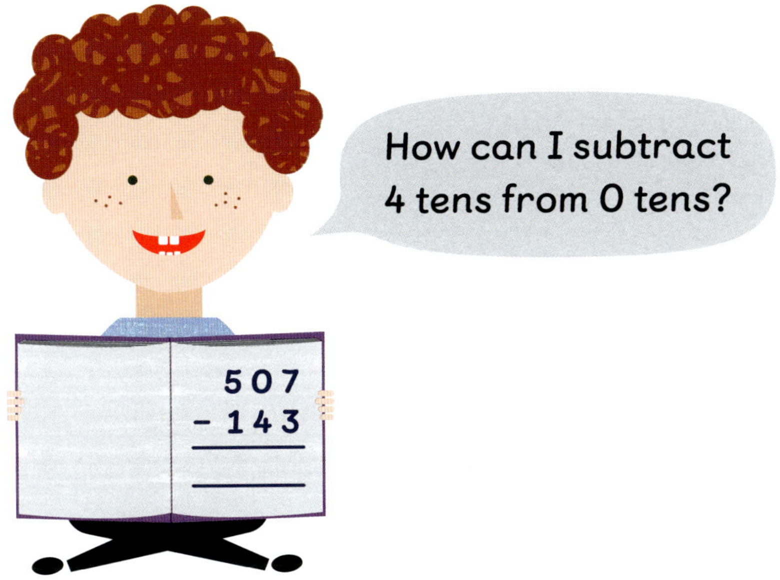

What does Elliott need to do to find the answer?

Master

507 – 143 = ▭

Step 1 Subtract the ones.

7 ones – 3 ones = 4 ones

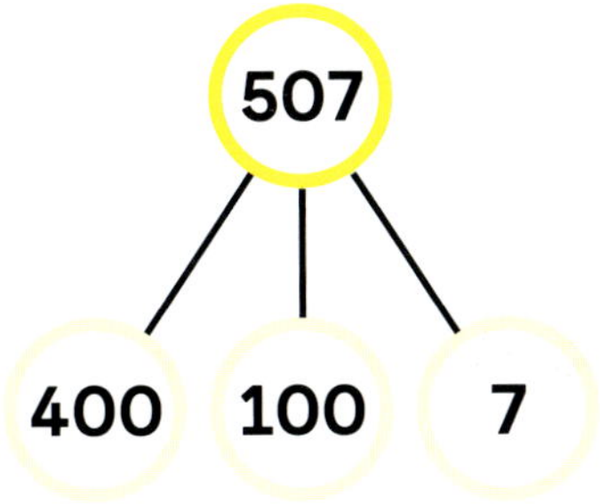

	h	t	o
	5	0	7
–	1	4	3
			4

Step 2 Rename 1 hundred as 10 tens.
Subtract the tens.

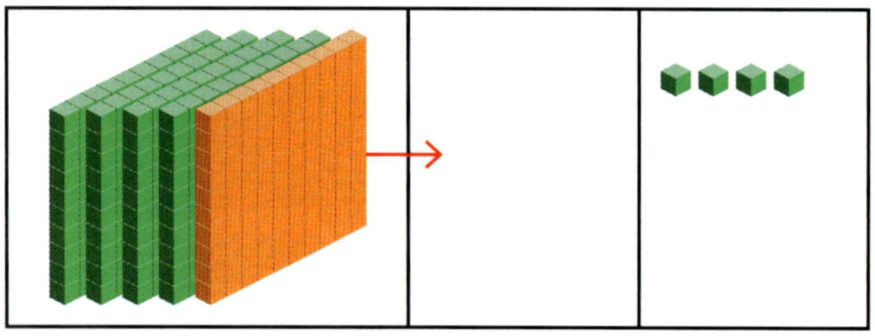

10 tens – 4 tens = 6 tens

	h	t	o
	4	10	
	~~5~~	~~0~~	7
–	1	4	3
		6	4

Step 3 Subtract the hundreds.

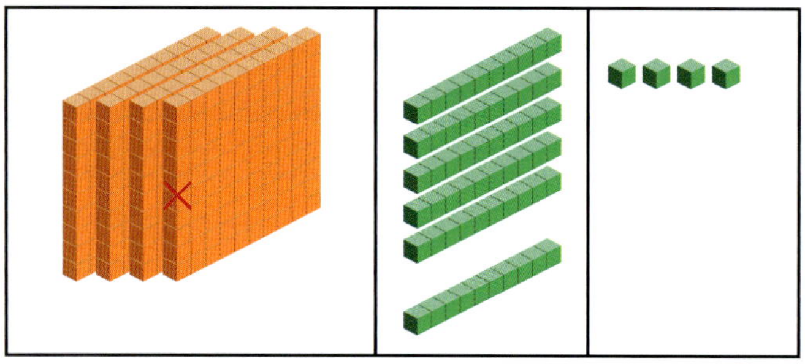

4 hundreds – 1 hundred = 3 hundreds

507 – 143 = 364

	h	t	o
	4	10	
	~~5~~	~~0~~	7
–	1	4	3
	3	6	4

Guided Practice

Subtract.

1. 849 – 57 =
2. 307 – 162 =
3. 463 – 281 =
4. 905 – 420 =

Complete Worksheet **16**: pages **55–56**

Subtracting with Renaming

Lesson 17

Explore

How can you find the answer to question 2?

1.

	7	3	0
−	4	1	0

2.

	7	3	0
−	4	6	7

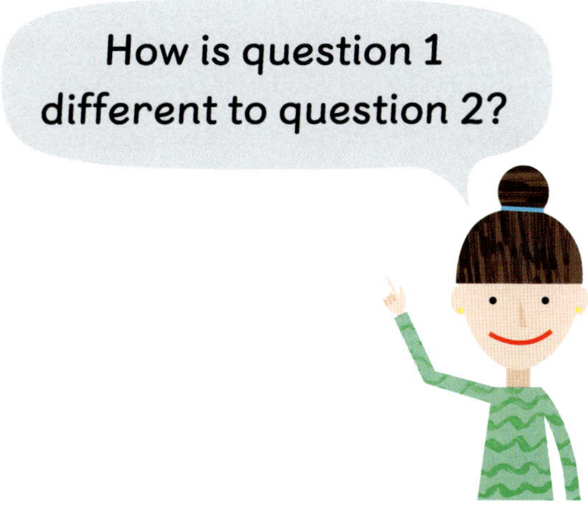

Master

730 − 467 =

Step 1 Rename 1 ten as 10 ones.
Subtract the ones.

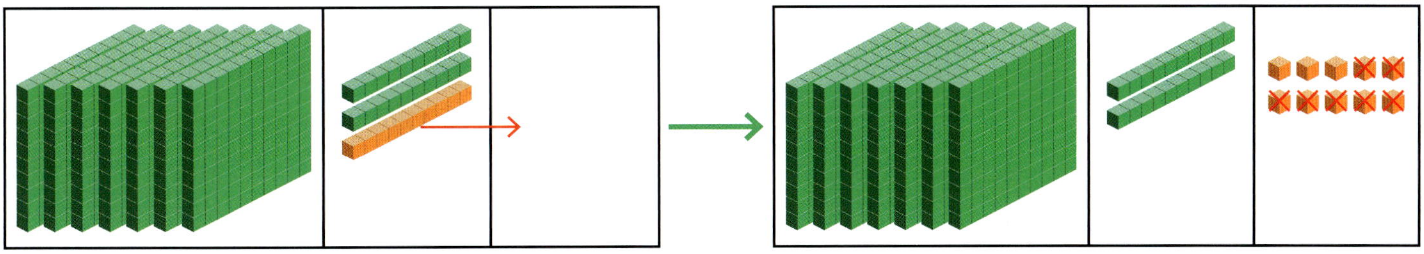

10 ones − 7 ones = 3 ones

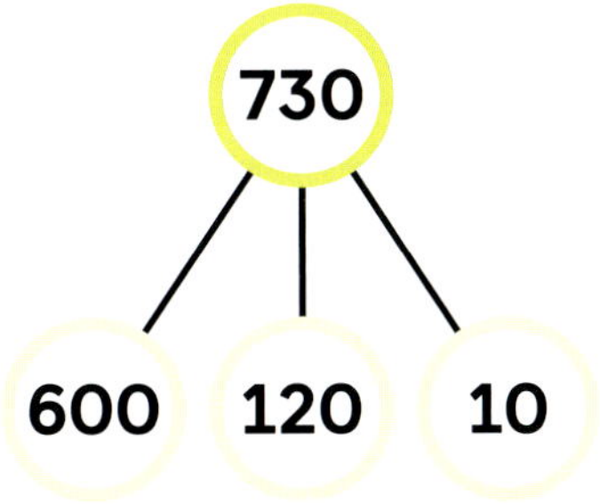

	h	t	o
		2	10
	7	~~3~~	~~0~~
−	4	6	7
			3

Step 2 Rename 1 hundred as 10 tens.
Subtract the tens.

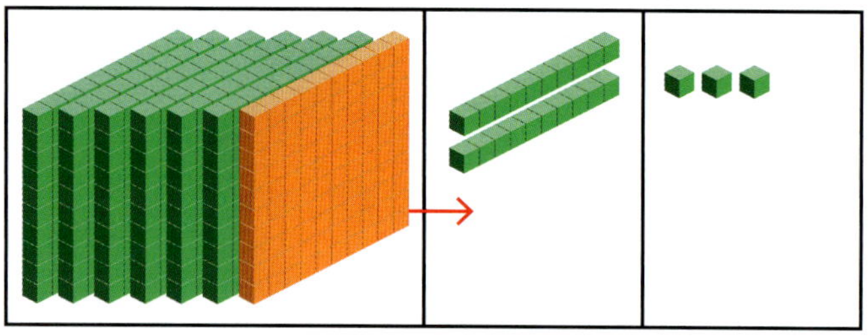

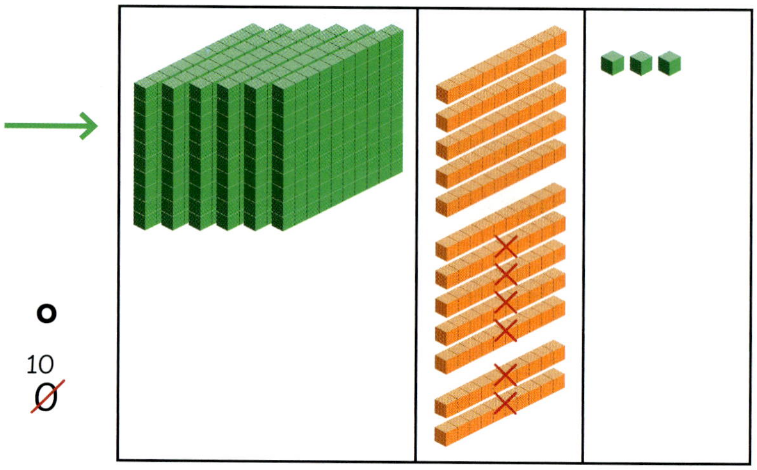

12 tens − 6 tens = 6 tens

	h	t	o
	6	12	10
	~~7~~	~~2~~ ~~3~~	~~0~~
−	4	6	7
		6	3

Step 3 Subtract the hundreds.

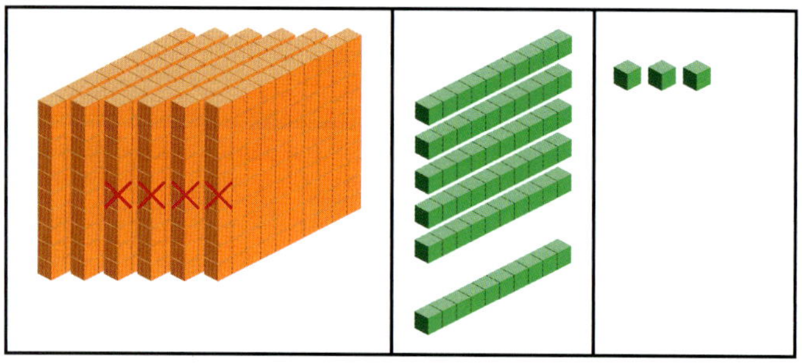

	h	t	o
	6	12	10
	~~7~~	~~2~~ ~~3~~	~~0~~
−	4	6	7
	2	6	3

6 hundreds − 4 hundreds = 2 hundreds

730 − 467 = 263

Check your answer using addition.
263 + 467 = 730

Guided Practice

Subtract.

1 346 − 68 = ☐

2 631 − 297 = ☐

3 720 − 96 = ☐

4 504 − 226 = ☐

Complete Worksheet **17**: pages **57–58**

Subtracting with Renaming

Lesson 18

Explore

There are 400 adults and children on an aeroplane.
164 of them are children.
How many adults are on the aeroplane?

Master

400 – 164 =

Step 1 Rename 1 hundred as 10 tens.

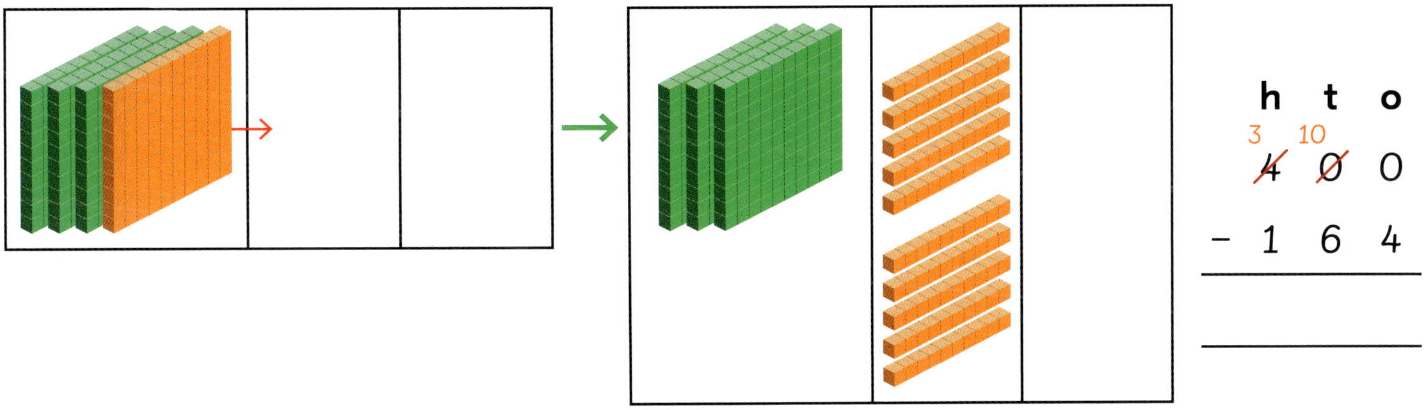

Step 2 Rename 1 ten as 10 ones.

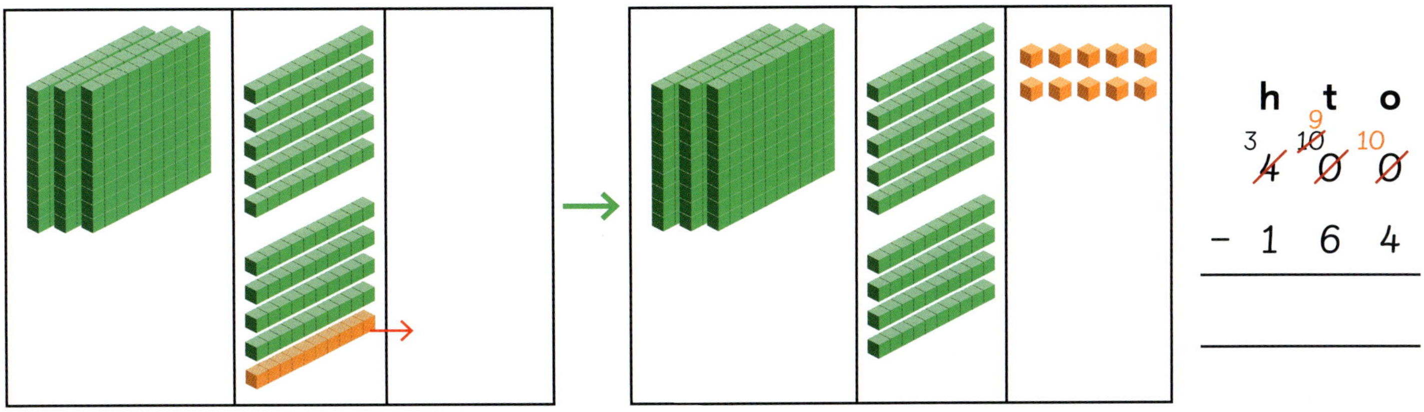

Step 3 Subtract the ones.
10 ones − 4 ones = 6 ones

	h	t	o
	3	9 ~~10~~	10
	~~4~~	~~0~~	~~0~~
−	1	6	4
			6

Step 4 Subtract the tens.
9 tens − 6 tens = 3 tens

	h	t	o
	3	9 ~~10~~	10
	~~4~~	~~0~~	~~0~~
−	1	6	4
		3	6

Step 5 Subtract the hundreds.
3 hundreds − 1 hundred = 2 hundreds

	h	t	o
	3	9 ~~10~~	10
	~~4~~	~~0~~	~~0~~
−	1	6	4
	2	3	6

400 − 164 = 236

There are 236 adults on the aeroplane.

Guided Practice

Subtract.

1. 200 − 45 = ☐
2. 500 − 68 = ☐
3. 400 − 234 = ☐
4. 900 − 432 = ☐

Complete Worksheet **18**: pages **59–60**

Using Models

Lesson 19

Explore

Hannah has 4 cupcakes.
Sam has 3 cupcakes.

Hannah　　　　Sam

How many cupcakes do Hannah and Sam have in total?

Master

1 Use [cubes] to show the cupcakes.

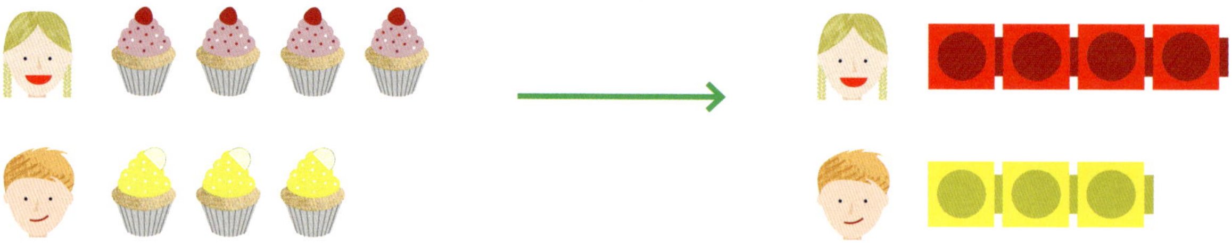

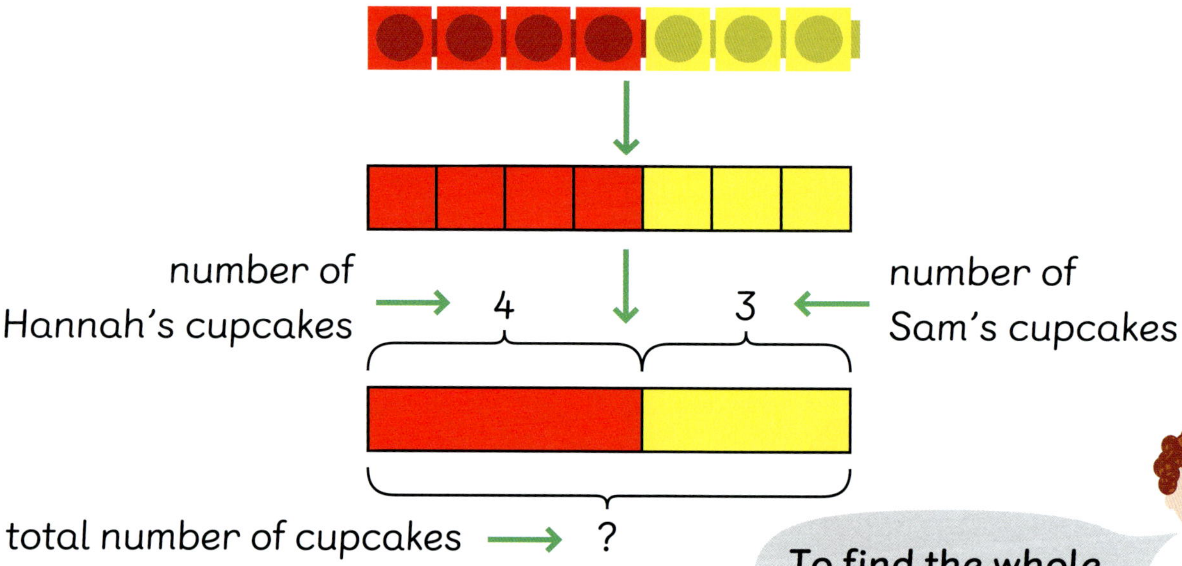

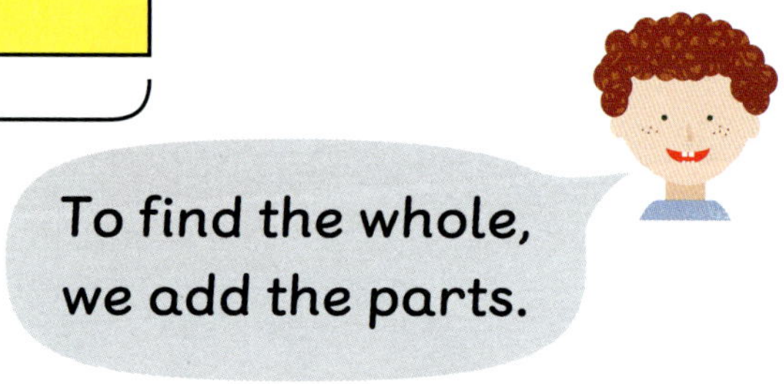

4 + 3 = 7 or 3 + 4 = 7

Hannah and Sam have 7 cupcakes in total.

2 [boy] has 17 marbles. [girl] has 21 marbles.

How many marbles do they have altogether?

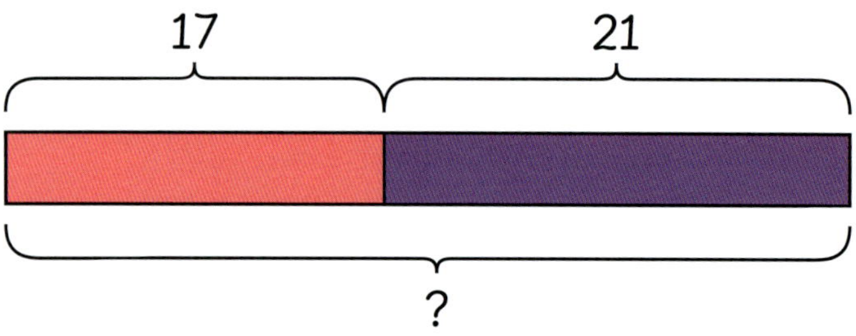

	t	o
	1	7
+	2	1
	3	8

17 + 21 = 38

They have 38 marbles altogether.

3 45 children are playing football and tennis.
27 children are playing football.
How many children are playing tennis?

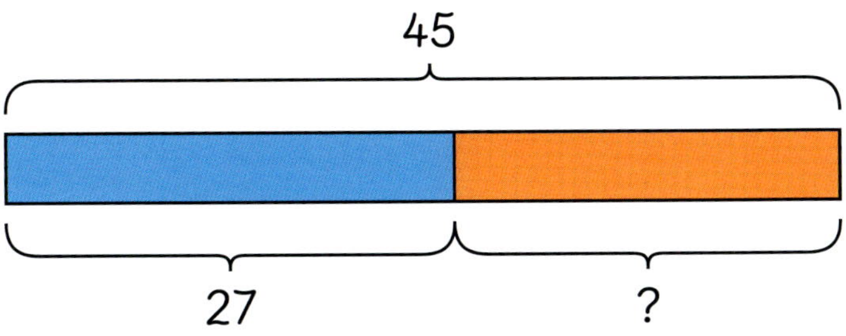

	t	o
	4	5
−	2	7
	☐	☐

☐ − 27 = ☐

☐ children are playing tennis.

To find a part, we subtract the other part from the whole.

Guided Practice

Solve.

1 Holly has 36 stickers.
Her dad gives her 27 more.
How many stickers does Holly have?

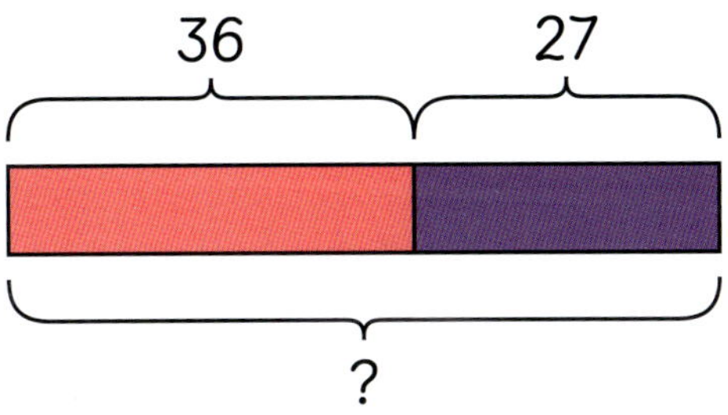

2 Oak has 63 stamps in a book.
39 are from England and the rest are from New Zealand.
How many stamps from New Zealand are in the book?

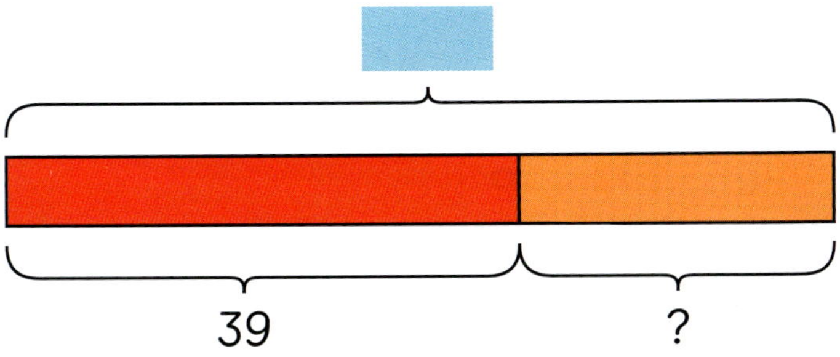

Complete Worksheet **19**: pages **61–62**

Using Models

Lesson 20

Explore

500 people are on a train.
360 of them are children.
How many of them are adults?

Master

We need to subtract 360 from 500.

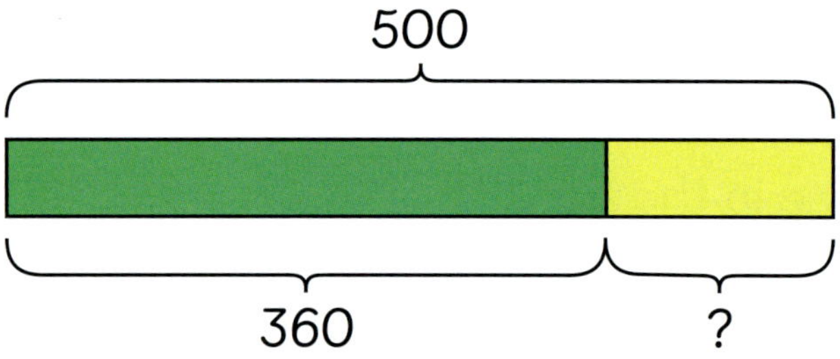

We can split 500 like this to subtract.

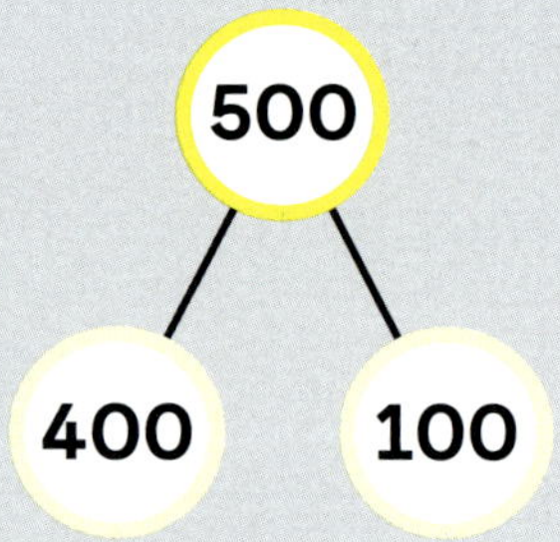

500 – 360 = 140

There are 140 adults on the train.

h	t	o
~~5~~ (4)	~~0~~ (10)	0
– 3	6	0
1	4	0

Guided Practice

Use models to solve the word problems.

1 A shop sold 38 jumpers.
It also sold 138 T-shirts.
How many items of clothing did the shop sell altogether?

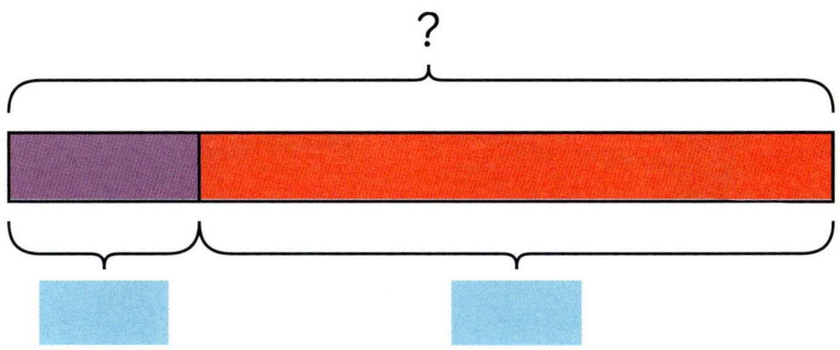

2 246 people travel on a train to Manchester.
327 people travel on a train to London.
How many people are travelling in total?

3 A builder has 400 nails.
He uses 102 of them.
How many nails does he have left?

Complete Worksheet **20**: pages **63–64**

Using Models

Lesson 21

Explore

How many more apples than oranges are there?

Master

1 There are 9 apples. There are 5 oranges.
Draw bars to show the number of apples and oranges.

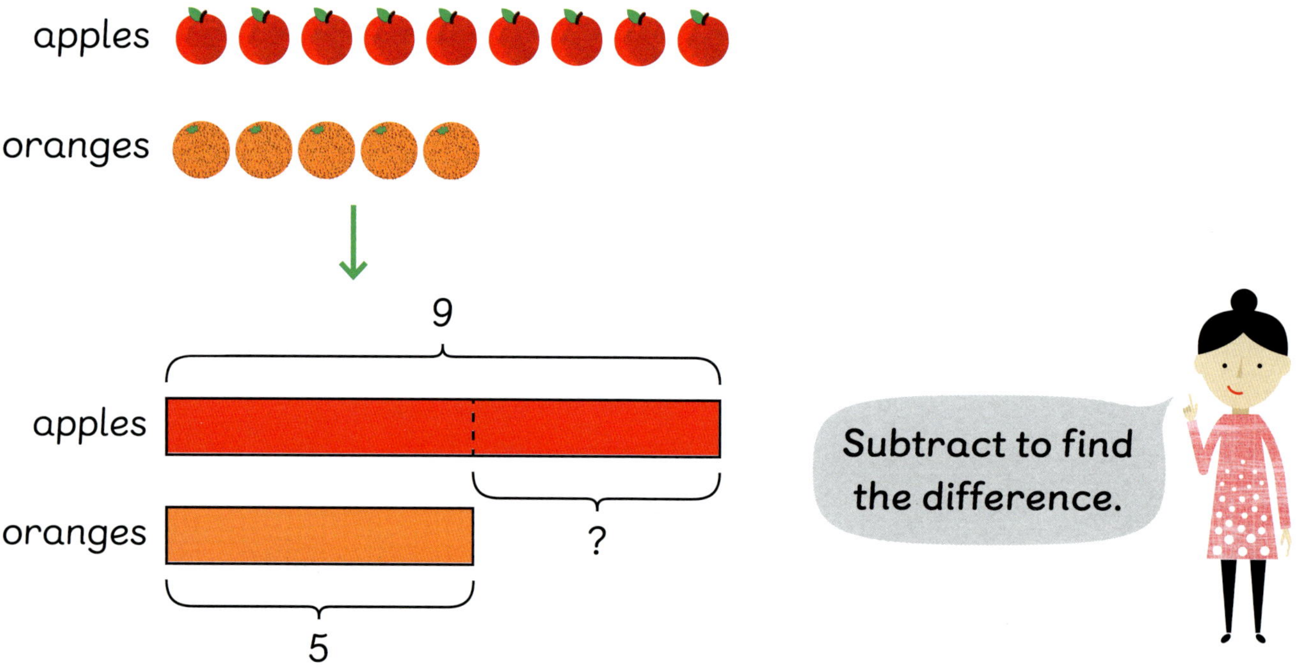

9 – 5 = 4

There are 4 more apples than oranges.

2 Charles has 47 chocolates. He has 19 more chocolates than Emma. How many chocolates does Emma have?

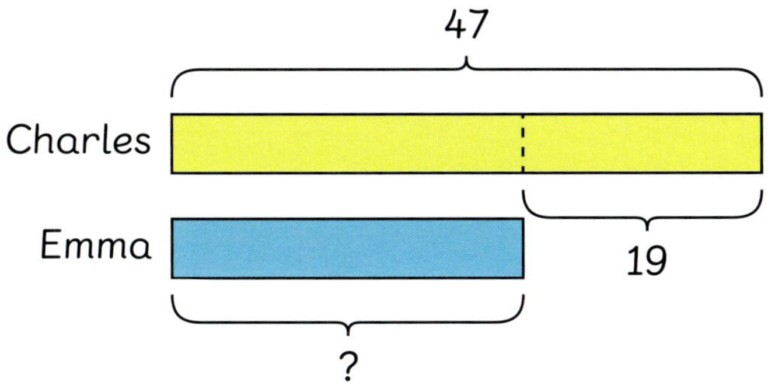

___ − ___ = ___

Emma has ___ chocolates.

Guided Practice

1 Sam made 18 paper aeroplanes. Holly made 12 paper aeroplanes. How many more paper aeroplanes did Sam make than Holly?

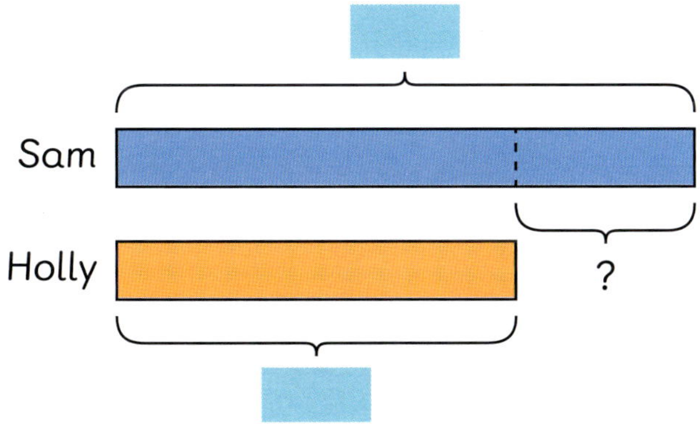

2 Elliott baked 57 buns. He baked 21 fewer buns than Lulu baked. How many buns did Lulu bake?

3 63 people visited a museum in the morning.
37 more people visited the museum in the afternoon than in the morning.
How many people visited the museum in the afternoon?

Complete Worksheet **21**: pages **65–66**

Using Models

Lesson
22

Explore

Ravi read 130 pages of his book.
Lulu read 84 pages of her book.
How many fewer pages did Lulu read than Ravi?

Master

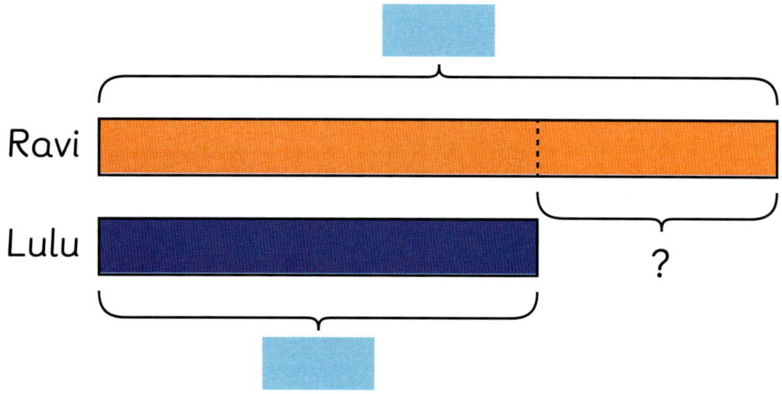

____ ◯ ____ = ____

Lulu read ____ fewer pages than Ravi.

Guided Practice

1 142 children are playing football.
56 fewer children are playing rugby.
How many children are playing rugby?

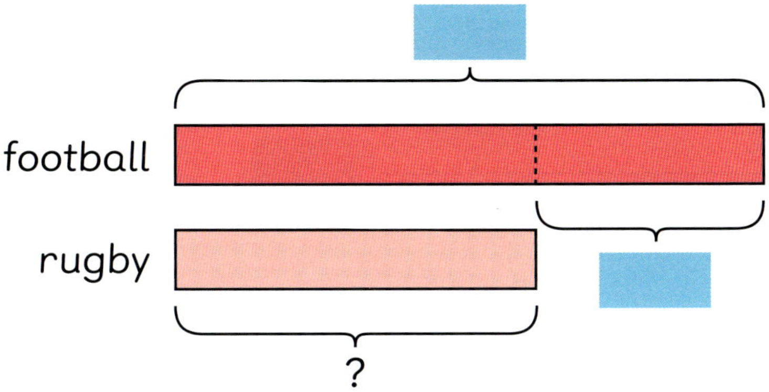

☐ ◯ ☐ = ☐

☐ children are playing rugby.

2 A farmer has 307 cows.
She has 60 more sheep than cows.
How many sheep does the farmer have?

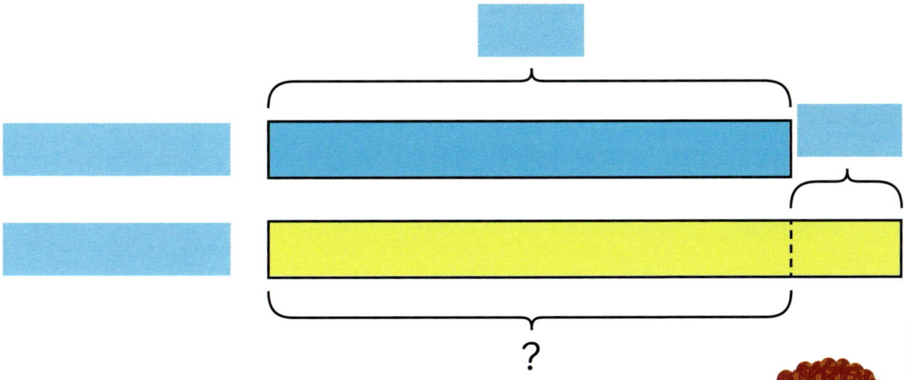

☐ ◯ ☐ = ☐

The farmer has ☐ sheep.

3 Charles picks 234 blackberries.
He picks 48 more blackberries than Holly picks.
How many blackberries does Holly pick?

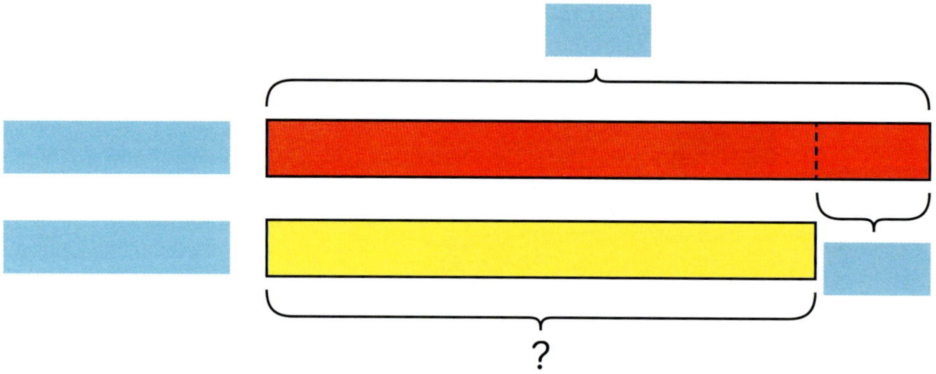

☐ ◯ ☐ = ☐

Holly picks ☐ blackberries.

4 The school library has 358 fiction books and 169 non-fiction books.
How many more fiction books than non-fiction books are in the library?

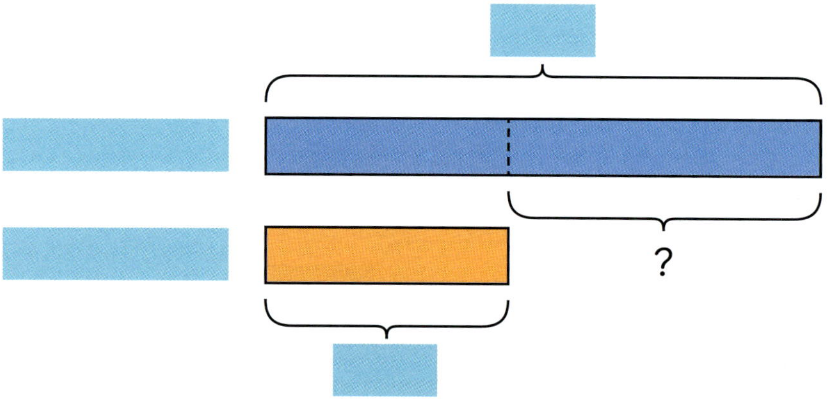

☐ ◯ ☐ = ☐

There are ☐ more fiction books than non-fiction books in the library.

Complete Worksheet **22**: pages **67–68**

Mind Challenge

Find the missing numbers below:

(a)

	6	□	2
+	□	9	□
	9	1	1

(b)

	□	□	5
−		8	□
	2	4	9

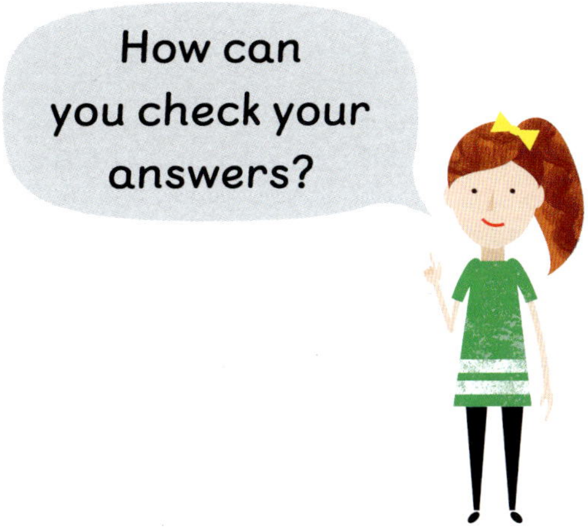

My Maths Journal

Use the numbers 310 and 270 to write an addition and subtraction story. You can also use words like **more** and **fewer** in your stories. Then, solve the problems in your stories using a model.

Review

I can...

- ☐ write an addition and subtraction fact family.
- ☐ add numbers up to 3 digits without renaming.
- ☐ add numbers up to 3 digits with renaming.
- ☐ subtract numbers up to 3 digits without renaming.
- ☐ subtract numbers up to 3 digits with renaming.
- ☐ use models to solve addition and subtraction problems, including those that require renaming.

Are there different ways we can count the number of mushrooms?

Chapter 3
Multiplication and Division

Multiplying by 3

Lesson 1

Explore

Emma made muffins for her friends.
How many muffins did she make in total?

Master

1 There are 3 muffins on each tray.

1 group of 3
$1 \times 3 = 3$

2 groups of 3
$2 \times 3 = 6$

3 groups of 3
$3 \times 3 = 9$

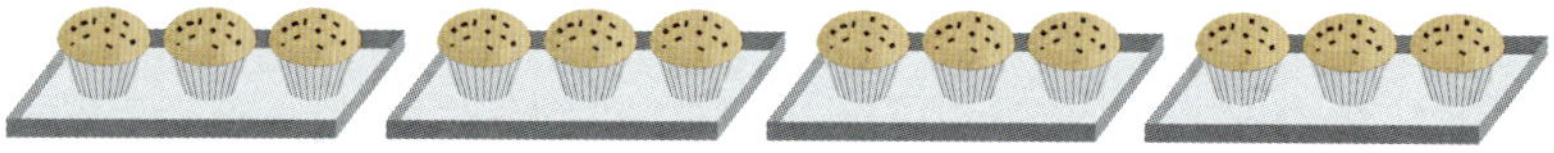

4 groups of 3
$4 \times 3 = 12$

5 groups of 3
$5 \times 3 = 15$

Emma made 15 muffins in total.

2

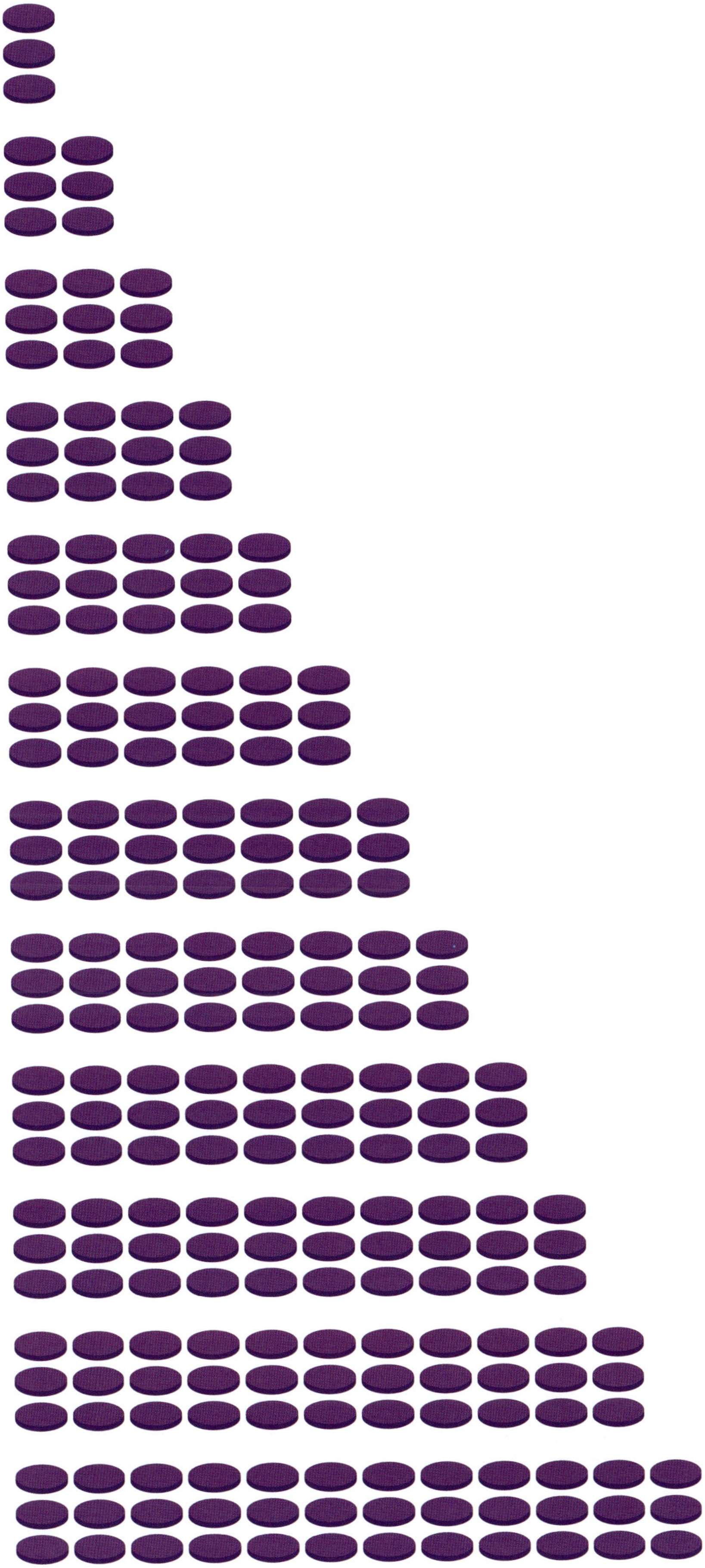

$1 \times 3 = 3$

$2 \times 3 = 6$

$3 \times 3 = 9$

$4 \times 3 = 12$

$5 \times 3 = 15$

$6 \times 3 = 18$

$7 \times 3 = 21$

$8 \times 3 = 24$

$9 \times 3 = 27$

$10 \times 3 = 30$

$11 \times 3 = 33$

$12 \times 3 = 36$

3 Count in threes.

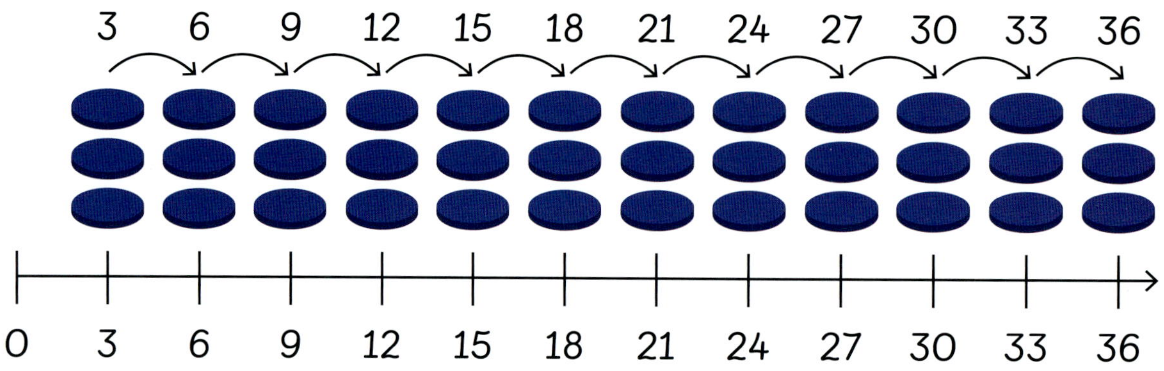

Activity Time

Play a game (2 pupils)

You will need:

1. Place all the  cards face up on the table.
2. Place the  cards face down in a pile.
3. Take turns to pick one  card and place it face up on the table.
4. The first pupil to find the  card with the same value keeps both cards.
5. Keep playing until there are no cards left.
6. The winner is the pupil with the most cards at the end.

Guided Practice

1

(a) $1 \times 3 =$

(b) $2 \times 3 =$

(c) $4 \times 3 =$

(d) $8 \times 3 =$

2

(a)

$3 \times 3 =$

$6 \times 3 =$

(b)

$5 \times 3 =$

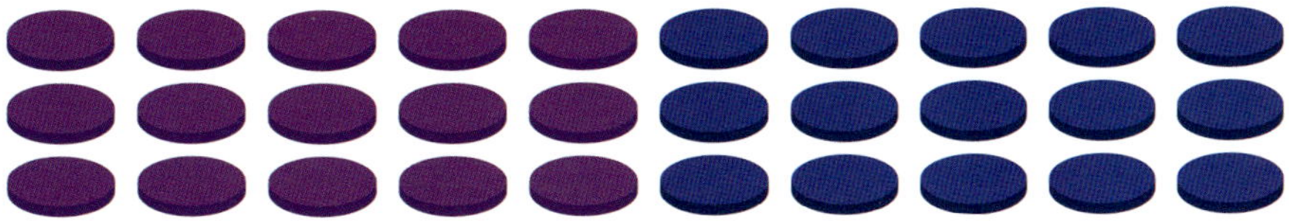

$10 \times 3 =$

Complete Worksheet **1**: pages **73–74**

Multiplying by 3

Lesson 2

Explore

How many flowers will be in the display when Ruby adds another row?

Master

1

There will be 15 flowers in the display.

2

Guided Practice

1 Find the missing numbers.

(a) 7 × 3 = 21

8 × 3 = 21 + ____ = ____

(b) 10 × 3 = ____

9 × 3 = ____ − 3 = ____

(c) 9 × 3 = ____

11 × 3 = ____ + 6 = ____

2 Fill in the blanks to complete the number patterns.

(a) 0, 3, 6, ____, ____, 15

(b) 12, ____, ____, 21, 24, ____

Complete Worksheet **2**: pages **75–76**

Multiplying by 4

Lesson 3

Explore

Use multiplication to find the total number of carrots.

Master

1

1 group of 4

$1 \times 4 = 4$

2 groups of 4

$2 \times 4 = 8$

3 groups of 4

$3 \times 4 = 12$

4 groups of 4

$4 \times 4 = 16$

5 groups of 4

$5 \times 4 = 20$

There are 20 carrots in total.

2

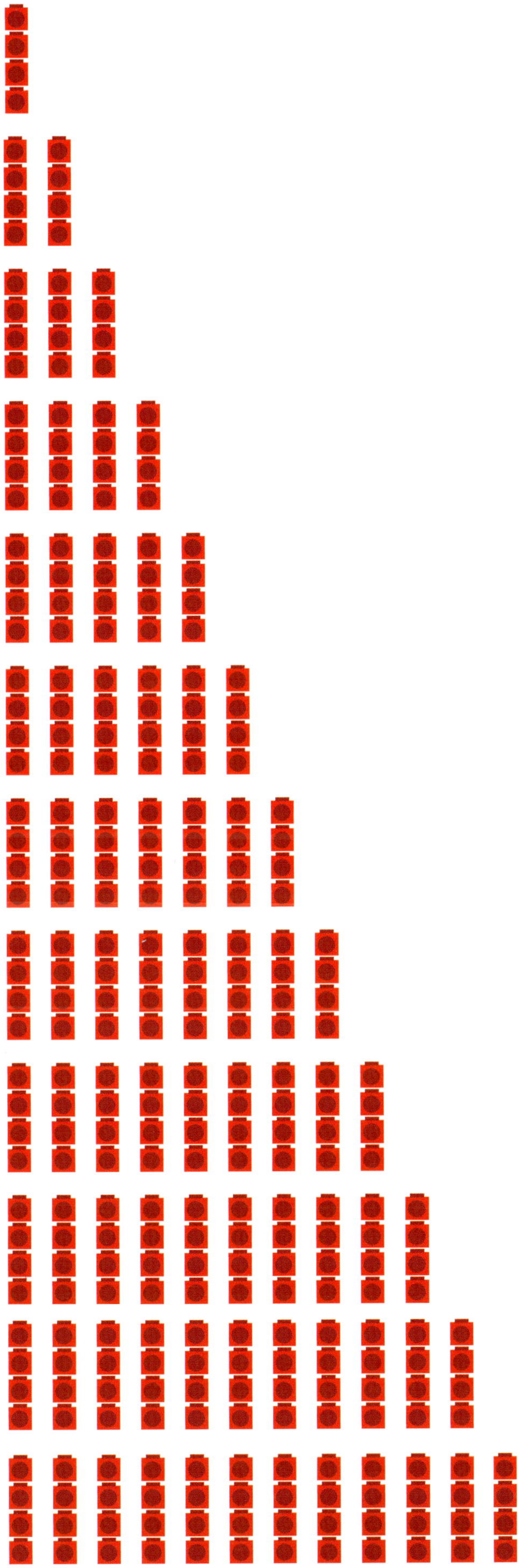

$1 \times 4 = 4$

$2 \times 4 = 8$

$3 \times 4 = 12$

$4 \times 4 = 16$

$5 \times 4 = 20$

$6 \times 4 = 24$

$7 \times 4 = 28$

$8 \times 4 = 32$

$9 \times 4 = 36$

$10 \times 4 = 40$

$11 \times 4 = 44$

$12 \times 4 = 48$

Guided Practice

Find the missing numbers.

1

____ × ____ = ____

2

(a) $1 \times 4 =$ ____ $2 \times 4 =$ ____

(b) $2 \times 4 =$ ____ $4 \times 4 =$ ____

(c) $3 \times 4 =$ ____ $6 \times 4 =$ ____

(d) $4 \times 4 =$ ____ $8 \times 4 =$ ____

(e) $5 \times 4 =$ ____ $10 \times 4 =$ ____

(f) $6 \times 4 =$ ____ $12 \times 4 =$ ____

3

(a)

$2 \times 4 =$ ____

(b)

$0 \times 4 =$ ____

Complete Worksheet **3**: pages **77–78**

Multiplying by 4

Lesson 4

Explore

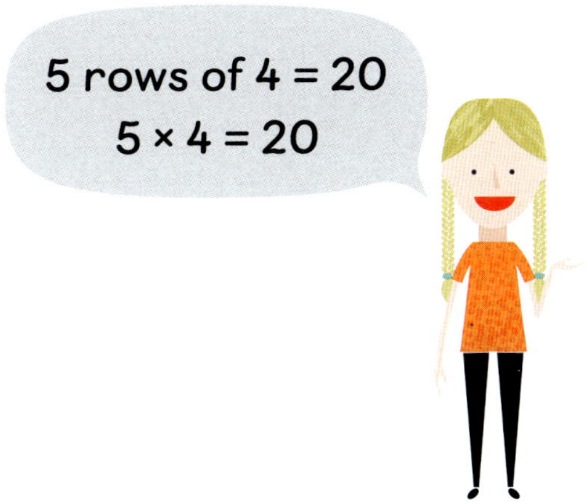

How does knowing that 5 × 4 = 20 help us find 4 × 4 or 6 × 4?

Master

1

2

3

$10 \times 4 = 40$

$9 \times 4 = 40 - \square = \square$

4

$9 \times 4 = 36$

$8 \times 4 = 36 - \square = \square$

$7 \times 4 = 36 - \square = \square$

Guided Practice

1 Find the missing numbers.

(a)

10 × 4 = 40

11 × 4 = ☐ + 4 = ☐

(b) 7 × 4 = ☐

6 × 4 = ☐ − 4 = ☐

(c) 10 × 4 = ☐

12 × 4 = ☐ + ☐ = ☐

2 (a) 6 × 4 = ☐

(b) 4 × 4 = ☐

(c) 2 × 4 = ☐

(d) 0 × 4 = ☐

3 Fill in the blanks to complete the number patterns.

(a) 0, 4, 8, ☐, ☐, 20, ☐

(b) 28, ☐, ☐, 40, ☐, 48

Complete Worksheet **4**: pages **79–80**

Multiplying by 4 and 8

Lesson 5

Explore

How many slices of cheese are there?

Master

1

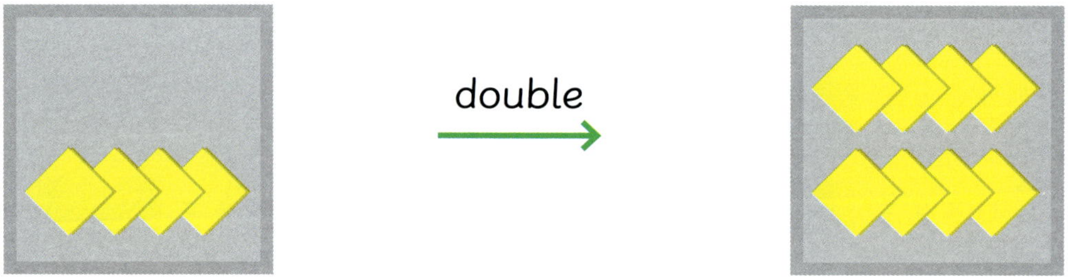

$1 \times 4 = 4$ $1 \times 8 = 8$

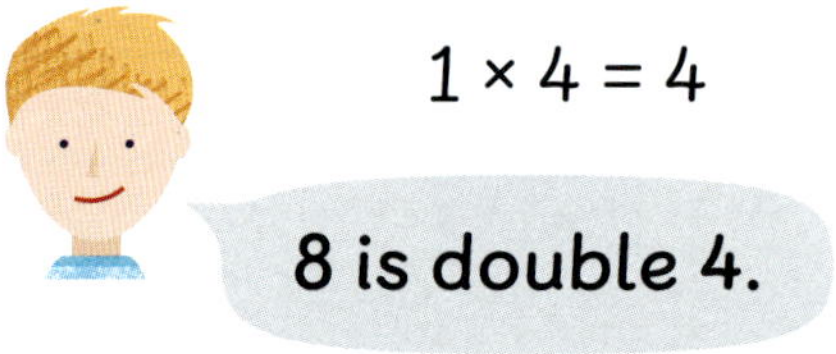

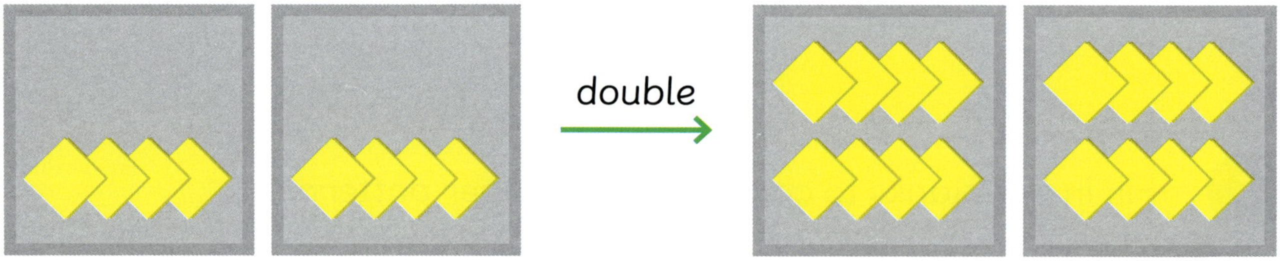

$2 \times 4 = 8$ $2 \times 8 = 16$

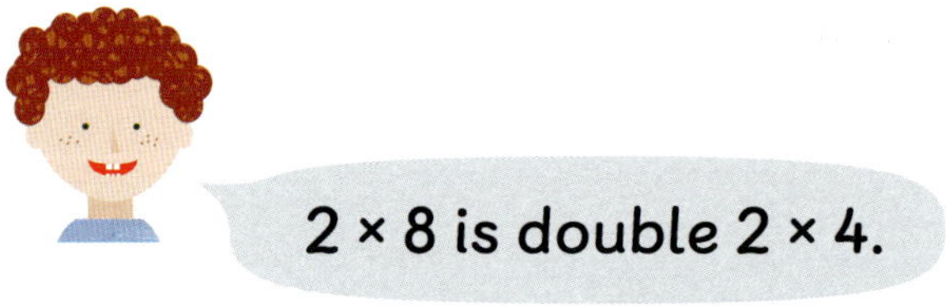

There are 16 slices of cheese.

2

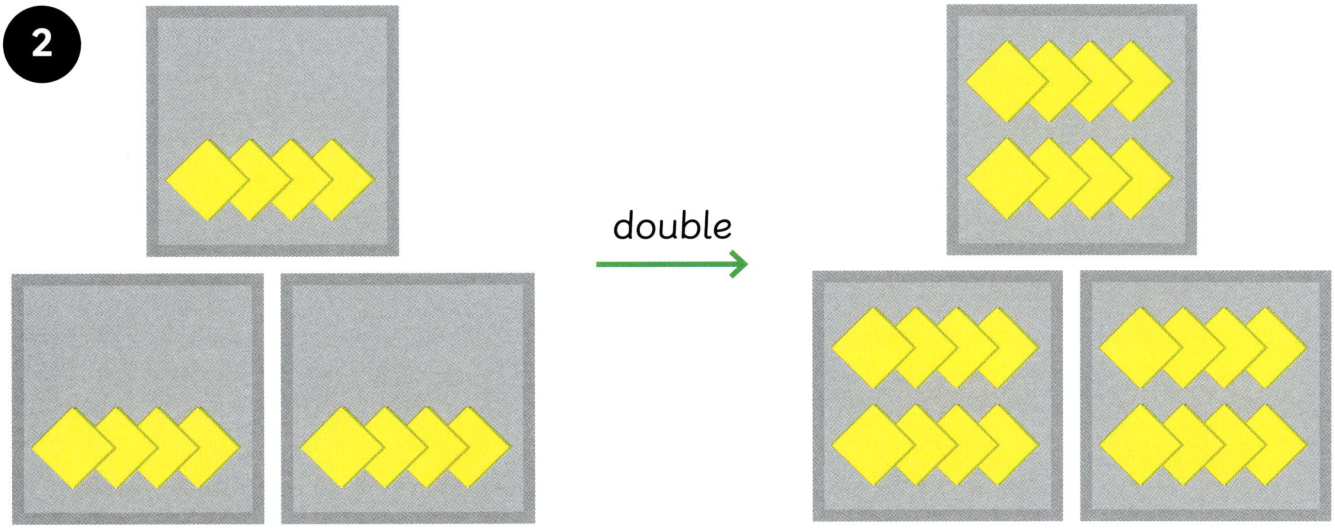

$3 \times 4 = 12$ $\qquad$ $3 \times 8 = 24$

3

$1 \times 4 = 4$	$1 \times 8 = 8$	$7 \times 4 = 28$	$7 \times 8 =$
$2 \times 4 = 8$	$2 \times 8 =$	$8 \times 4 = 32$	$8 \times 8 =$
$3 \times 4 = 12$	$3 \times 8 =$	$9 \times 4 = 36$	$9 \times 8 =$
$4 \times 4 = 16$	$4 \times 8 =$	$10 \times 4 = 40$	$10 \times 8 =$
$5 \times 4 = 20$	$5 \times 8 =$	$11 \times 4 = 44$	$11 \times 8 =$
$6 \times 4 = 24$	$6 \times 8 =$	$12 \times 4 = 48$	$12 \times 8 =$

Guided Practice

1 Multiply.

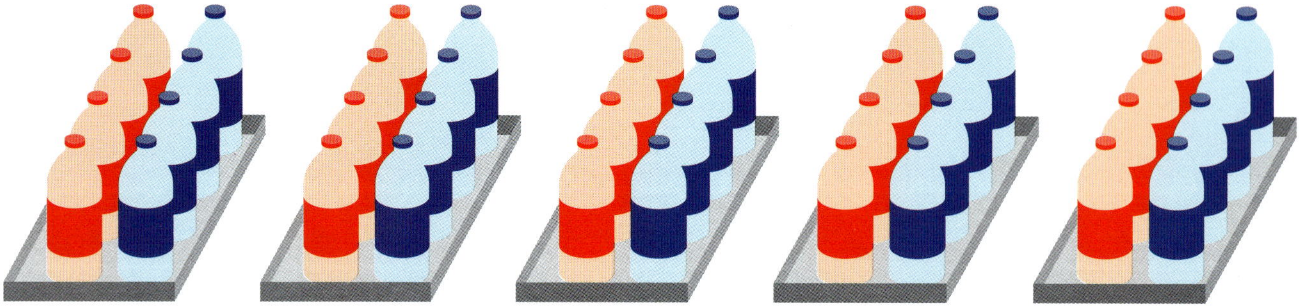

(a) 2 × 4 =

2 × 8 =

(b) 5 × 4 =

5 × 8 =

(c) 6 × 4 =

3 × 8 =

(d) 6 × 8 =

7 × 8 =

8 × 8 =

2 Find the product of:

(a) 3 and 4

(b) 4 and 8

(c) 4 and 6

Complete Worksheet **5**: pages **81–82**

Lesson 6

Multiplying by 8

Explore

How many mushrooms are there?

Master

1

There are 5 rows of 8 mushrooms.
5 × 8 = 40

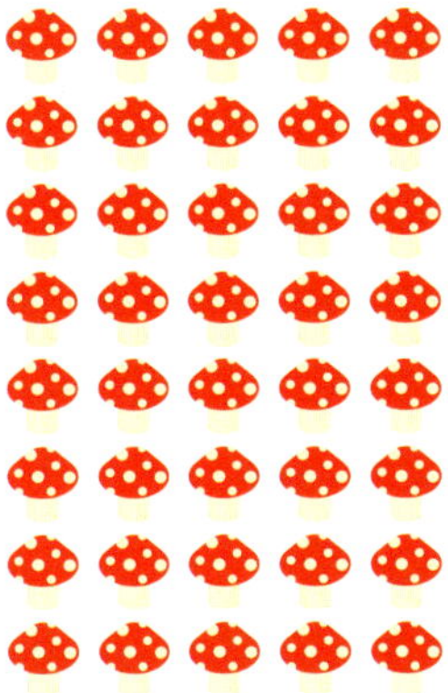

There are 8 rows of 5 mushrooms.
8 × 5 = 40

There are 40 mushrooms.

2

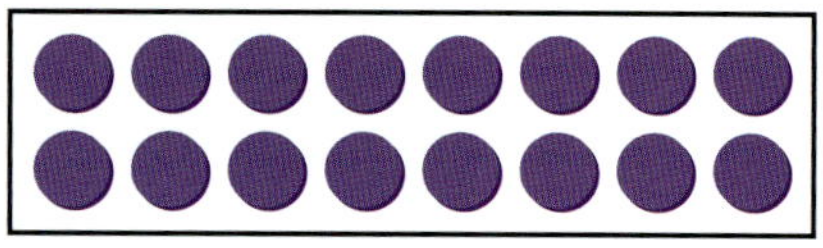

$2 \times 8 = 16$

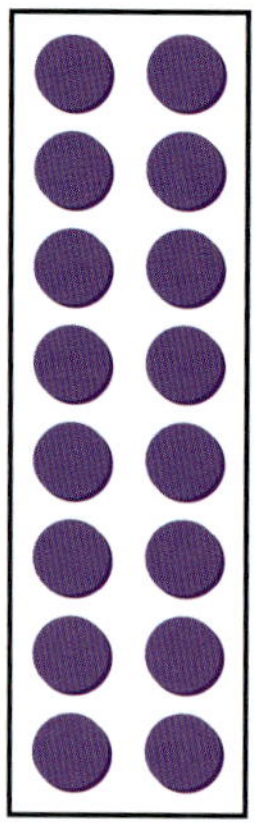

$8 \times 2 = 16$

The product of 2 and 8 is 16.
The product of 8 and 2 is also 16.

Guided Practice

Lulu is learning how to find the product of two numbers.
Can you help her by filling in the blanks?

1 (a) 5 and 4 ______ (b) 4 and 5 ______

(c) 5 and 8 ______ (d) 8 and 5 ______

2 (a) 4 and 7 ______ (b) 7 and 4 ______

(c) 8 and 7 ______ (d) 7 and 8 ______

Complete Worksheet **6**: pages **83–84**

Multiplying by 8

Lesson 7

Explore

How many tentacles do the octopuses have altogether?

Master

1 Each octopus has 8 tentacles.

1 group of 8

$1 \times 8 = 8$

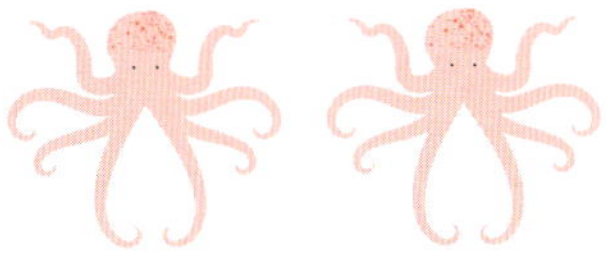

2 groups of 8

$2 \times 8 = 16$

3 groups of 8

$3 \times 8 = 24$

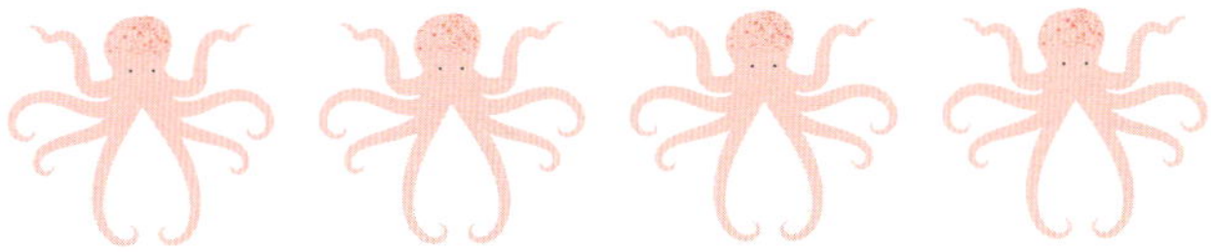

4 groups of 8

$4 \times 8 = 32$

The octopuses have 32 tentacles altogether.

2 Count in eights.

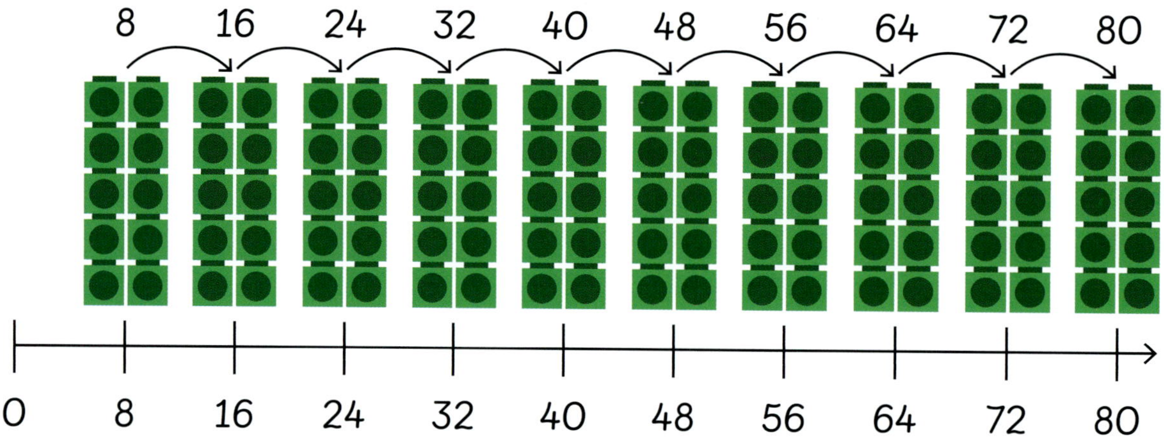

Guided Practice

1 Hannah uses 8 beads for each bracelet.
She makes 4 bracelets.
How many beads does Hannah use?

___ × ___ = ___

Hannah uses ___ beads.

2 Fill in the blanks to complete the number patterns.

(a)

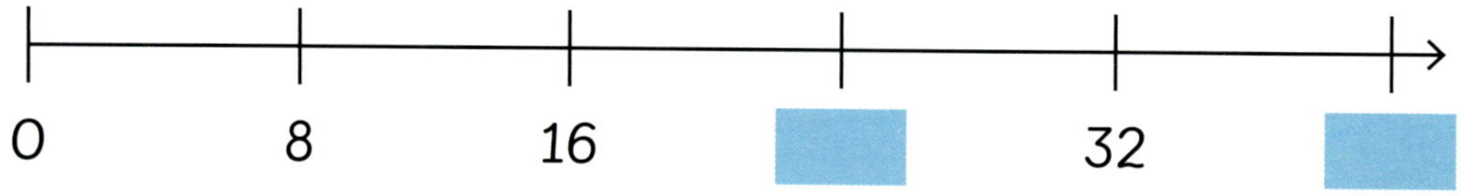

(b)

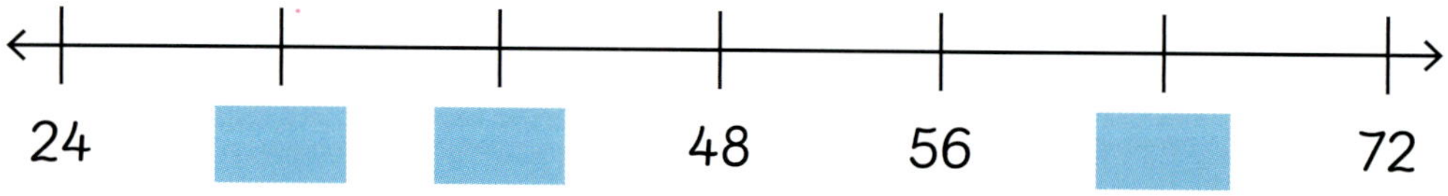

3

$2 \times 8 =$ ☐

$2 \times 8 =$ ☐

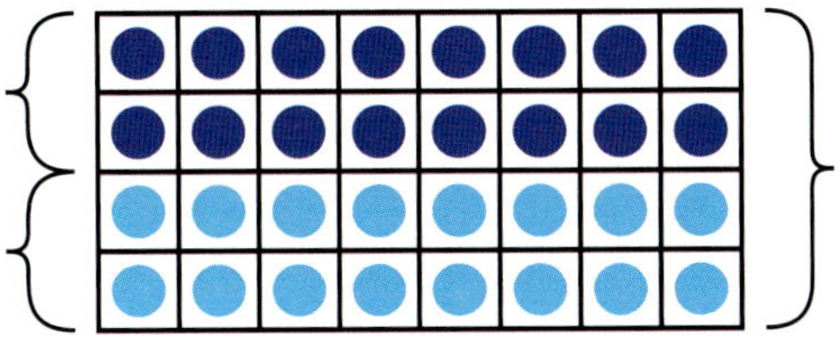

$4 \times 8 =$ ☐

$4 \times 8 =$ ☐

$1 \times 8 =$ ☐

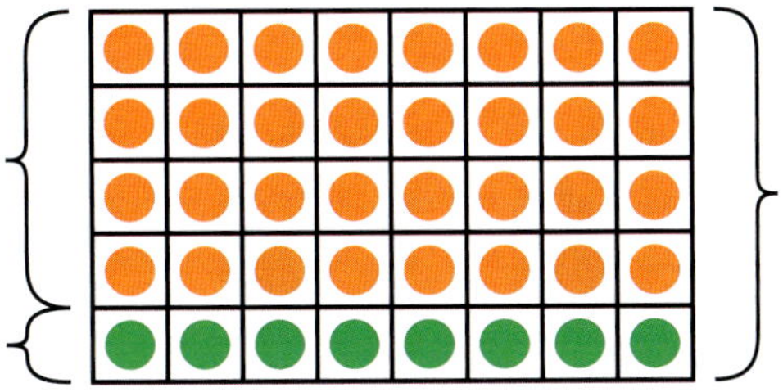

$5 \times 8 =$ ☐

$3 \times 8 =$ ☐

$3 \times 8 =$ ☐

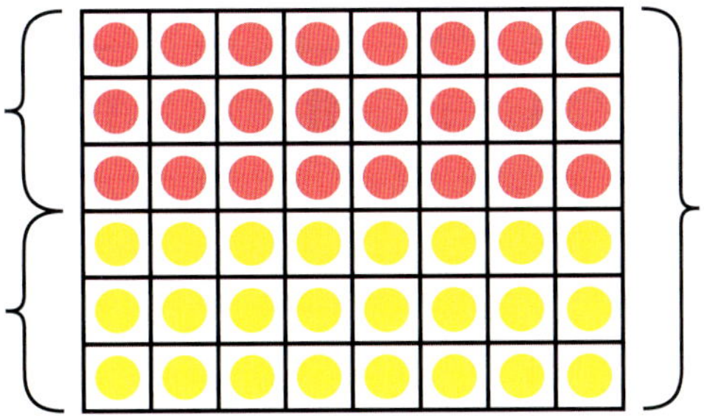

$6 \times 8 =$ ☐

$10 \times 8 =$ ☐

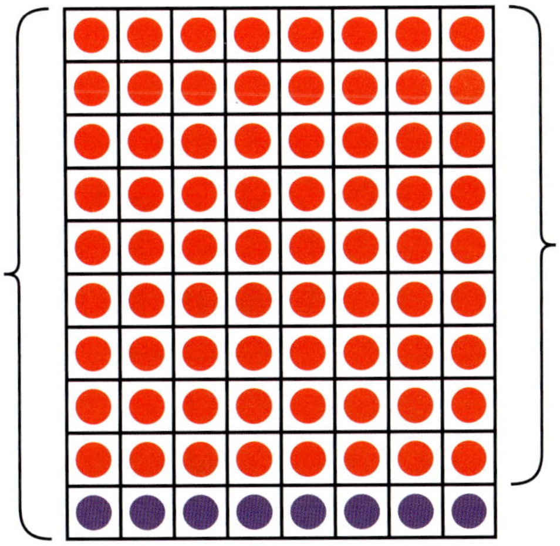

$9 \times 8 =$ ☐

Complete Worksheet **7**: pages **85–86**

Dividing by 3

Lesson 8

Explore

12 children are going sailing. There are 5 sailboats available.
Each sailboat can fit up to 4 children.
How many sailboats will the children use to go sailing?

Master

1 Put the children into groups of 3.

$12 \div 3 = 4$

If the children sail in groups of 3, they will use 4 sailboats.

2 Put 12 children into 3 equal groups.

$12 \div 3 = 4$

If the children are put into 3 equal groups, there will be 4 children in each sailboat. They will use 3 sailboats.

Guided Practice

Use  and  to help you.

1 (a) Put 18  into groups of 3.

There are ______ groups.

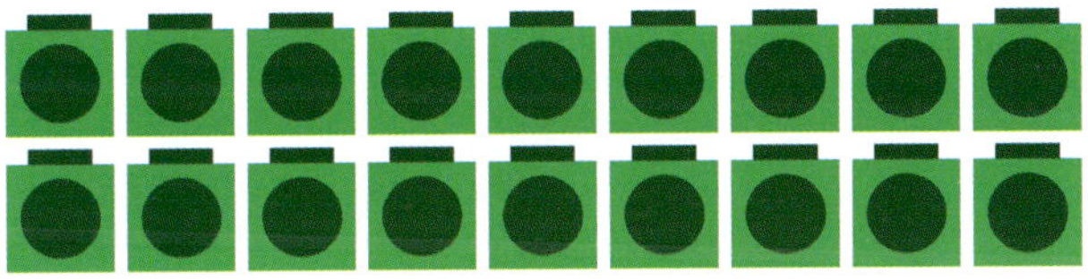

(b) Put 18  into 3 equal groups.

There are ______  in each group.

2 Fill in the missing numbers.

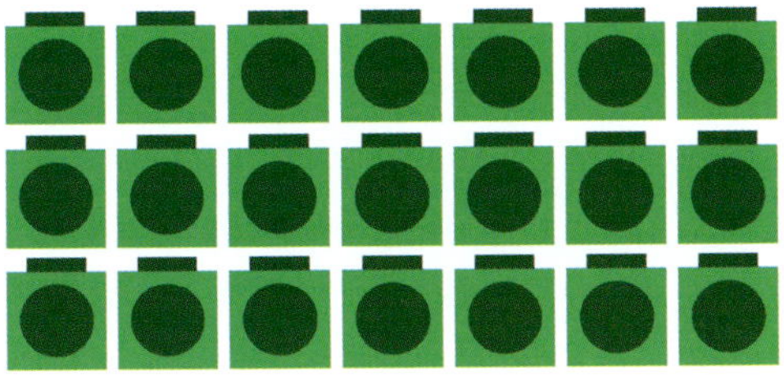

(a) $21 \div 3 =$ ______

$3 \times$ ______ $= 21$

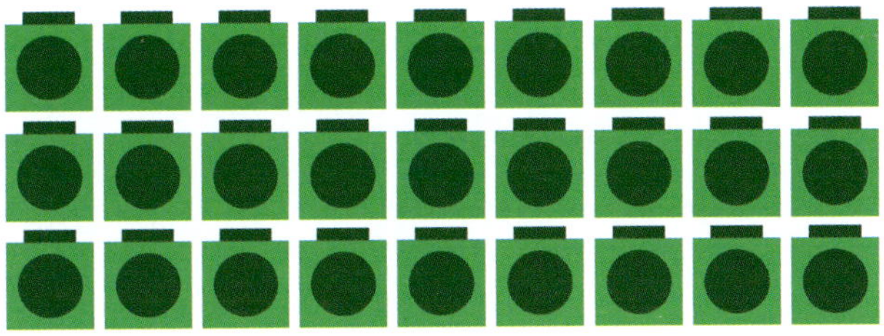

(b) $27 \div 3 =$ ______

______ $\times 3 = 27$

Complete Worksheet **8**: pages **87–88**

Dividing by 4

Lesson 9

Explore

Holly gives the tennis balls equally to 4 friends.

How many does each friend receive?

Master

1

$8 \div 4 = 2$

Each friend receives 2 .

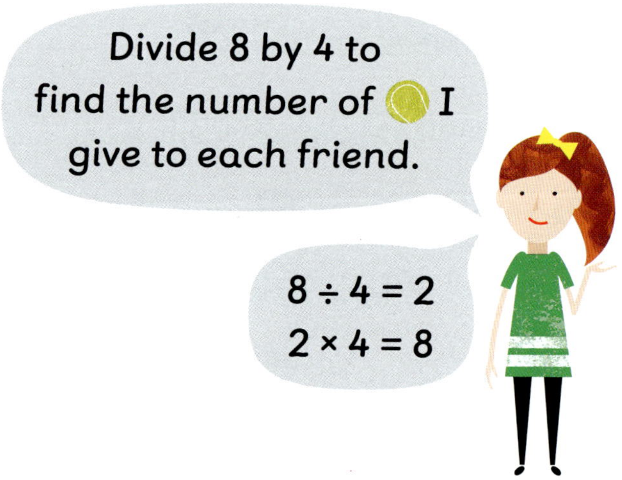

2 Holly puts the balls away in bags.
If she puts 4 in each bag,
how many bags will she use?

$8 \div 4 = 2$

Holly will use 2 bags.

Guided Practice

Use cubes and plates to help you.

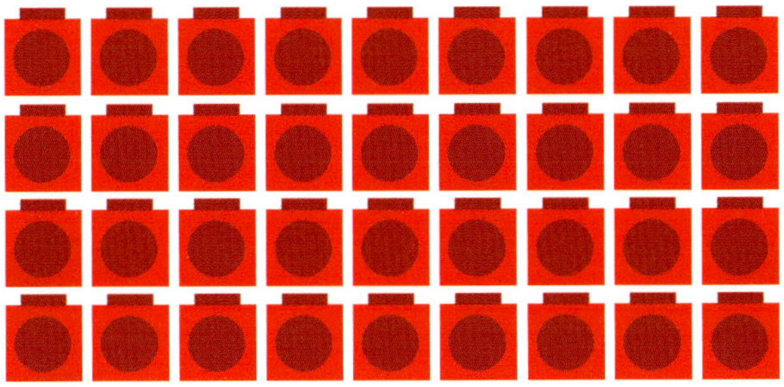

1 (a) Put 12 cubes into groups of 4.

There are ___ groups.

(b) Put 12 cubes into 4 equal groups.

There are ___ cubes in each group.

2 (a) Put 36 cubes into groups of 4.

There are ___ groups.

(b) Put 36 cubes into 4 equal groups.

There are ___ cubes in each group.

3

(a) $40 \div 4 =$ ___

$4 \times$ ___ $=$ ___

There are ___ groups.

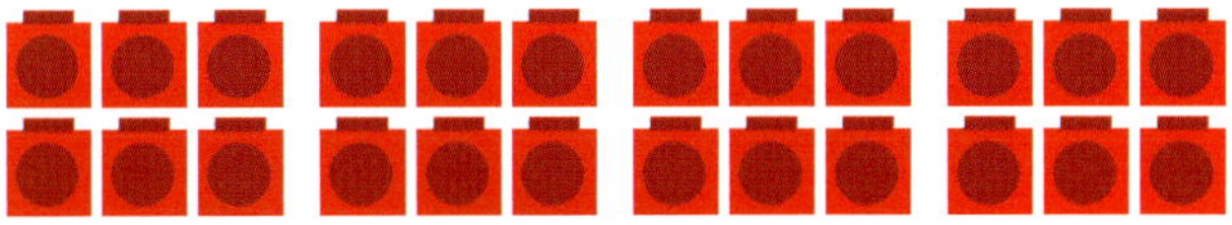

(b) $24 \div 4 =$ ___

$4 \times$ ___ $=$ ___

There are ___ in each group.

Complete Worksheet **9**: pages **89–90**

Multiplying and Dividing

Lesson 10

Explore

Amira and Ruby are making pizzas.
They have 12 olives.
They want to put 3 or 4 olives on each pizza.
Can we make a family of multiplication and division equations to help them?

Master

1

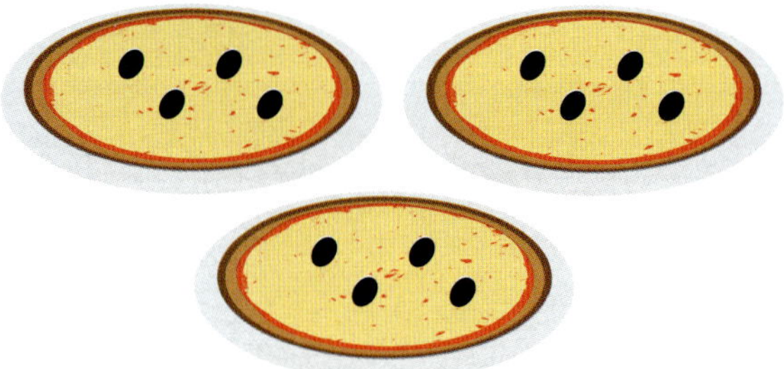

$12 \div 3 = 4$
$4 \times 3 = 12$

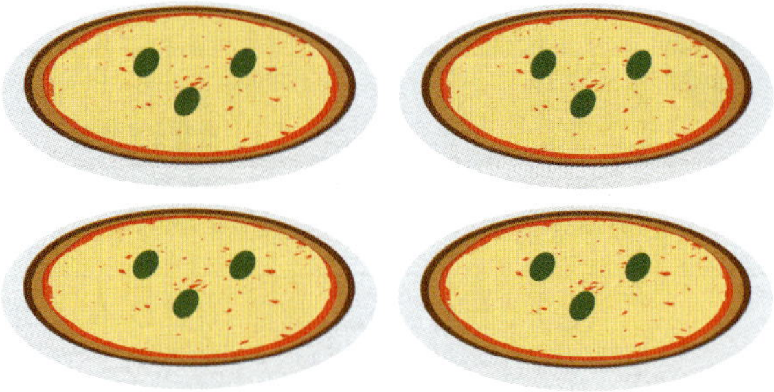

$12 \div 4 = 3$
$3 \times 4 = 12$

2

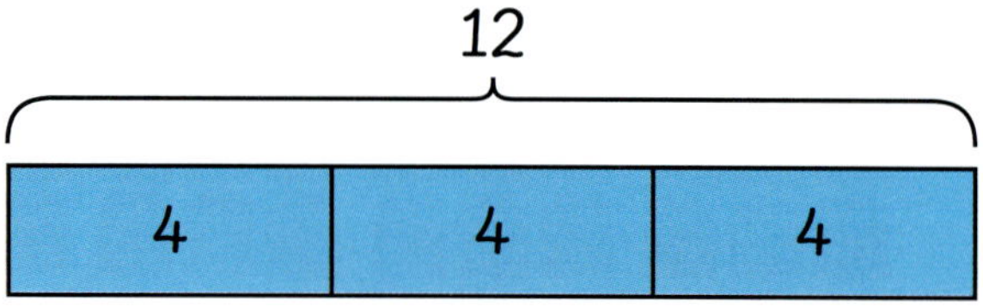

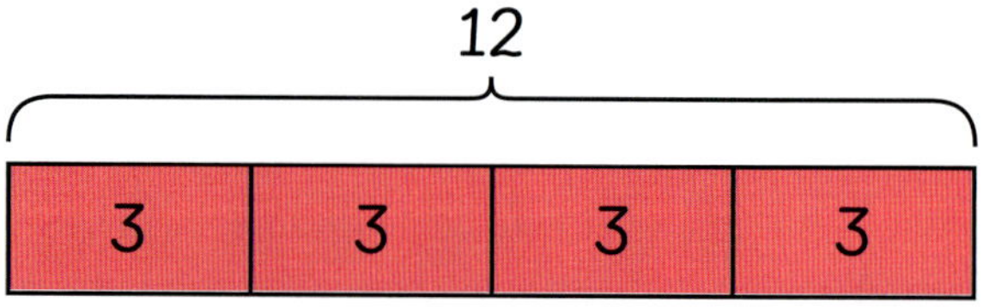

4 times 3 is 12, so 12 divided by 3 is 4.

12 divided into groups of 4 is equal to 3.

3 times 4 is 12, so 12 divided by 4 is 3.

12 shared between 4 is equal to 3.

Activity Time

Partner work

You will need:

1. Roll three dice and add together the three numbers you roll.
2. Pick up the same number of counters as the sum of the three numbers.
3. Arrange the counters into equal groups. If this is not possible, roll one of the dice again.
4. Write the multiplication and division fact family.
5. Check each other's answers.
6. Repeat the activity.

$3 \times 4 = 12$

$4 \times 3 = 12$

$12 \div 3 = 4$

$12 \div 4 = 3$

$2 \times 6 = 12$

$6 \times 2 = 12$

$12 \div 2 = 6$

$12 \div 6 = 2$

Guided Practice

Complete the multiplication and division fact family.

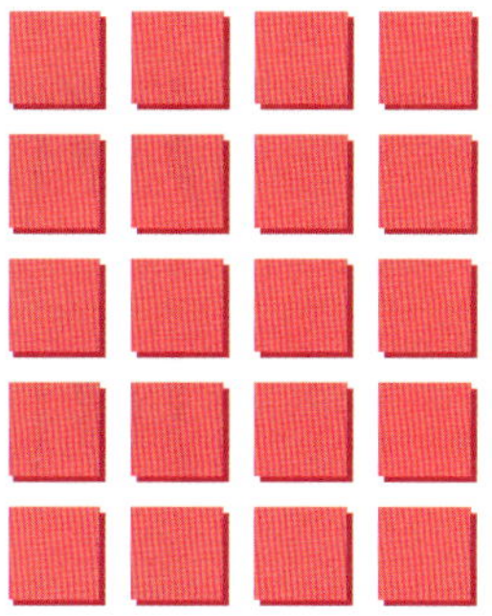

5 × 4 = ☐   ☐ ÷ ☐ = ☐

☐ × ☐ = ☐   ☐ ÷ ☐ = ☐

Complete Worksheet **10**: pages **91–92**

Dividing by 4 and 8

Lesson 11

Explore

Sam is helping his mum pack cobs of corn to sell at the market. How can he pack the cobs of corn into 4 and 8 equal groups?

Master

1 Sam put 32 cobs of corn into 4 equal groups.

$32 \div 4 = 8$
Each group has 8 cobs of corn.

2 Sam put 32 cobs of corn into 8 equal groups.

$32 \div 8 = 4$
Each group has 4 cobs of corn.

Guided Practice

1 Solve.

(a) Ravi puts the pears equally into 8 boxes. How many pears are in each box?

There are ___ pears in each box.

(b) He makes boxes of 4 bananas. How many boxes can he make?

He can make ___ boxes.

2 (a) $16 \div 8 =$ ___ ___ $\times 8 = 16$

$16 \div 4 =$ ___ ___ $\times 4 = 16$

(b) $48 \div 4 =$ ___ ___ $\times 4 = 48$

$48 \div 8 =$ ___ ___ $\times 8 = 48$

Complete Worksheet **11**: pages **93–94**

Solving Word Problems

Lesson 12

Explore

Sam planted some rows of seedlings. Can you use multiplication to find out the number of seedlings he planted?

Master

1 There are 3 seedlings in each row.

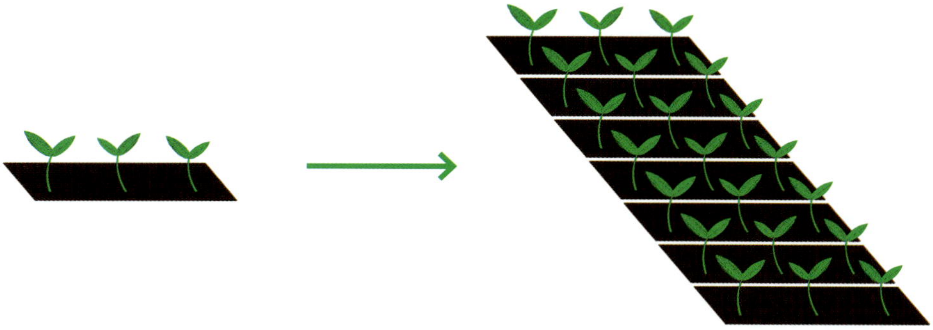

There are 7 rows of 3 seedlings.

$7 \times 3 = 21$

Sam planted 21 seedlings.

2 Amira planted 4 rows of seedlings.
Each row has 6 seedlings.
How many seedlings has she planted?

$4 \times 6 = 24$

Amira planted 24 seedlings.

Guided Practice

Solve the word problems.

1 A tricycle has 3 wheels.
How many wheels are on 6 tricycles?

☐ × ☐ = ☐

2 One car has 4 wheels.
How many wheels do 5 cars have?

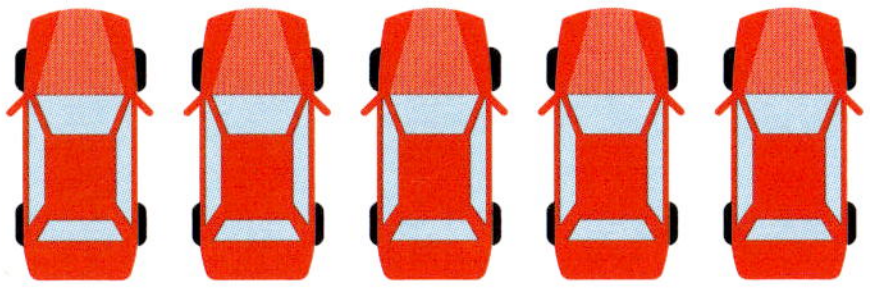

☐ × ☐ = ☐

3 There are 7 chairs.
Each chair has 4 legs.
How many legs are there altogether?

4 Each envelope contains 3 sheets of paper.
How many sheets of paper are in 5 envelopes?

5 Hannah has 3 pencil cases with 6 pencils in each of them.
Charles has 2 pencil cases with 8 pencils in each of them.
How many pencils do Hannah and Charles have altogether?

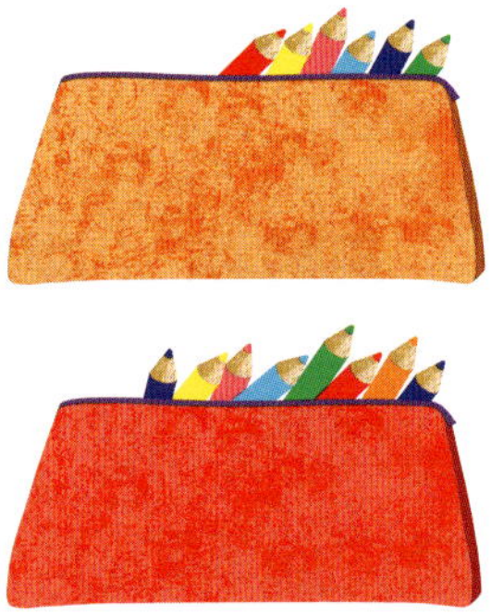

Complete Worksheet **12**: pages **95–96**

Solving Word Problems

Lesson 13

Explore

How many mushrooms has Lulu collected?
How can she put them equally onto plates?

Master

1 Lulu put the mushrooms into groups of 4.
Can you help her multiply to find the total number of mushrooms she collected?

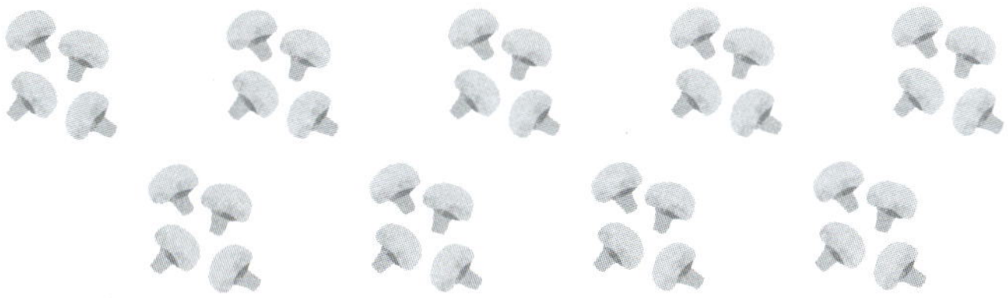

There are 4 mushrooms in each group.
There are 9 groups.

$9 \times 4 = 36$

Lulu collected 36 mushrooms.

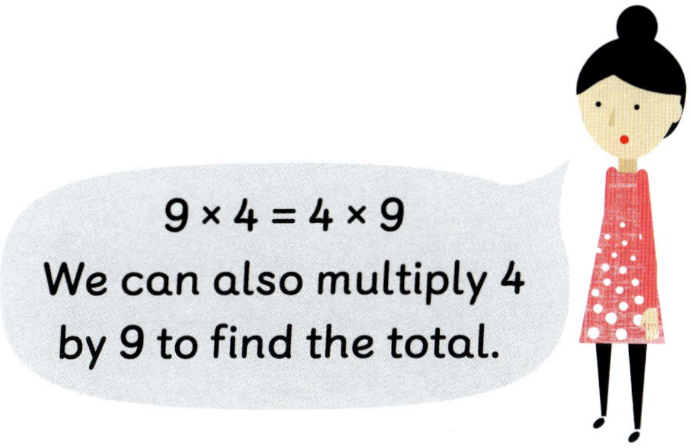

2 There are 36 mushrooms.
Lulu put 4 mushrooms on each plate.
Divide to find the number of plates of mushrooms.

$36 \div 4 = 9$

Lulu put the mushrooms equally onto 9 plates.

3 There are 3 boats.
24 children get into the boats.
Each boat has the same number of children.
How many children are in each boat?

___ ◯ ___ = ___

There are ___ children in each boat.

Guided Practice

Solve the word problems.

1 Ravi collects 27 apples from 3 apple trees.
He collects the same number of apples from each tree.
How many apples does he collect from each tree?

2 32 children are camping in tents.
There are 4 children in each tent.
How many tents are there?

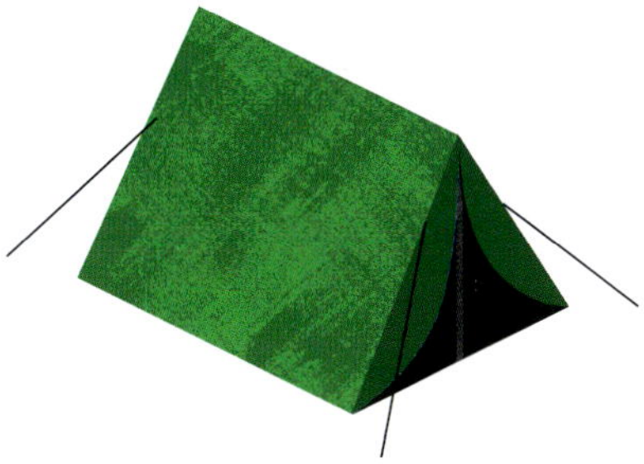

3 There are 48 people in a room.
An equal number of people sit at each of the 4 tables in the room.
How many people sit at each table?

4 Elliott's school buys 56 bulbs to plant in the garden.
They plant the bulbs in groups of 8.
How many groups of 8 bulbs can be planted?

Complete Worksheet **13**: pages **97–98**

Solving Word Problems

Lesson 14

Explore

How many hens are in the red hen house?

Master

1

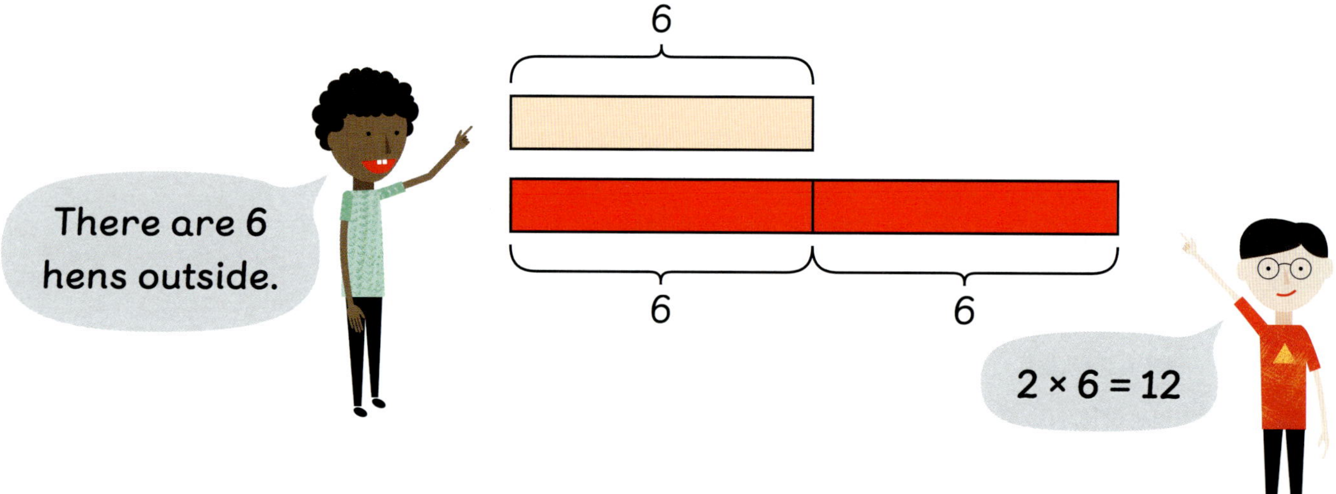

There are 12 hens in the red hen house.

2

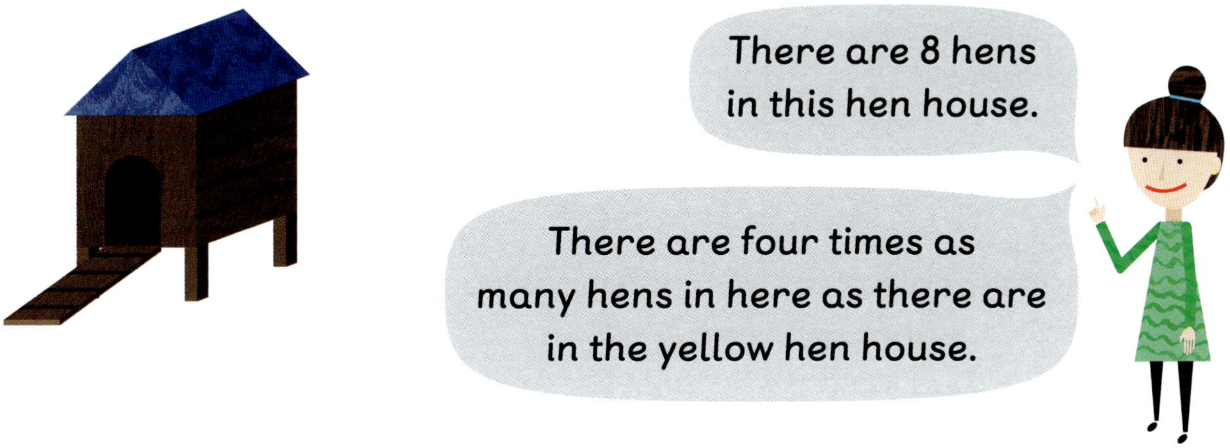

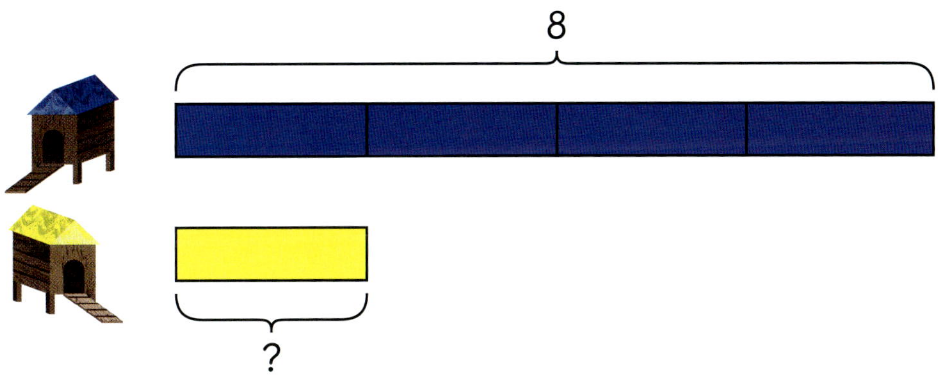

$8 \div 4 = 2$

 has 2 hens in it.

Guided Practice

Solve the word problems.

1  has saved 8 times as much money as  has saved.

 has saved £40.

How much money has  saved?

2 There are 8 pens in 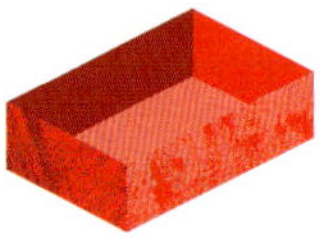.

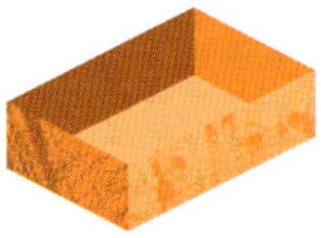 has 4 times as many pens in it as 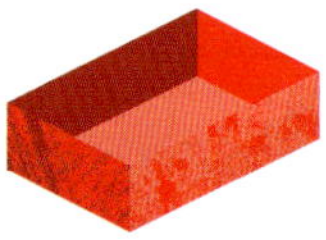 has.

 has twice as many pens in it as  has.

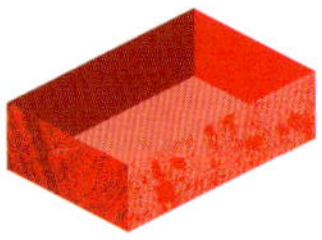 has ☐ pens.

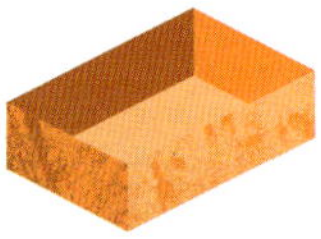 has ☐ pens.

 has ☐ pens.

3 There are 6 pears in a shop. There are 4 times as many apples as pears in the shop. The number of apples in the shop is 3 times the number of oranges. How many oranges are there in the shop?

Complete Worksheet **14**: pages **99–100**

Solving Word Problems

Lesson 15

Explore

Ruby has 3 hats and 2 scarves. How many different ways can she wear them?

Master

Ruby can wear the yellow scarf or the blue scarf.

There are 3 hats Ruby can wear with each scarf.

There are 6 different ways in total.

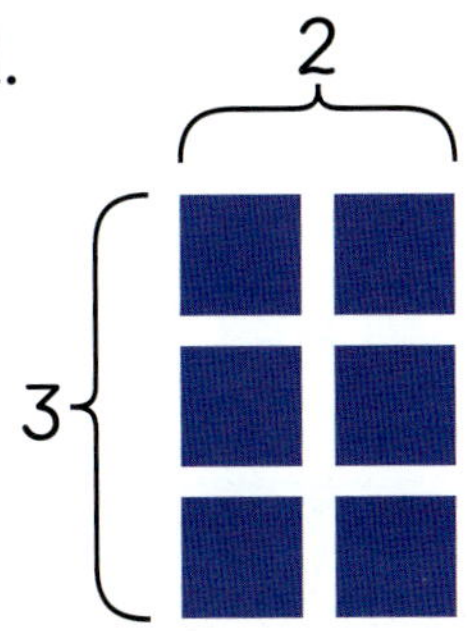

$3 \times 2 = 6$

Guided Practice

Solve the word problems.

1 Pick an ice-cream flavour and a topping.

vanilla strawberry chocolate

chocolate chips sprinkles syrup

How many different combinations of ice-cream flavours and toppings can you make?

2 Pick a bread roll.

white brown seeded

Pick a filling.

cheese jam cucumber peanut butter

How many different rolls can be made?

Complete Worksheet **15**: pages **101–102**

Mind Challenge

Sam, Emma and Ruby have some coins.
They have fewer than 50 coins altogether.

How many coins could the children have?

My Maths Journal

What has Ravi noticed about multiples of 3?
What has Holly noticed about multiples of 4?
Are they both correct?

What do you notice about multiples of 8?

Review

I can...

- ☐ recall and use the 3 times table.
- ☐ recall and use the 4 times table.
- ☐ recall and use the 8 times table.
- ☐ multiply and divide a number by 3.
- ☐ multiply and divide a number by 4.
- ☐ multiply and divide a number by 8.
- ☐ solve word problems involving multiplication and division using the 3, 4 and 8 times tables.

Chapter 4
Further Multiplication and Division

Multiplying 2-Digit Numbers

Lesson
1

Explore

How many counters are there altogether?

Master

1 Multiply 2 ones by 4.

4 × 2 ones = 8 ones
4 × 2 = 8

2 Multiply 2 tens by 4.

4 × 2 tens = 8 tens
4 × 20 = 80

There are 80 counters altogether.

8 tens = 80

4 × 20 = 20 × 4

Guided Practice

Multiply.

1 $4 \times 6 = \square$

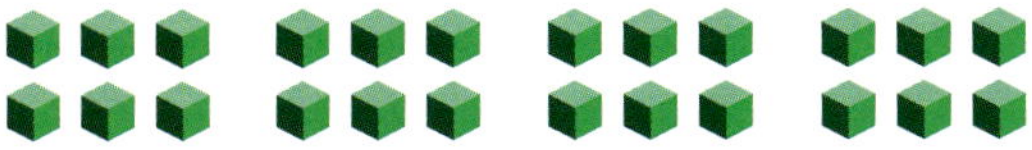

$4 \times 60 = \square$

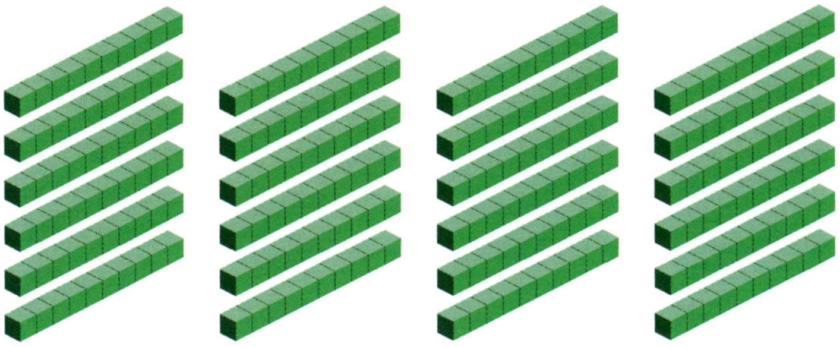

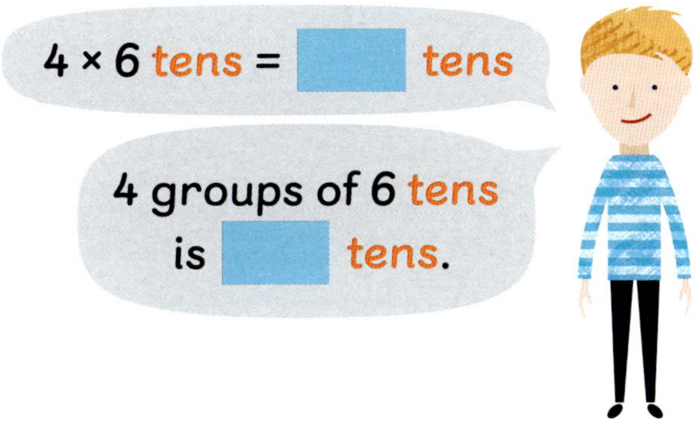

2 $70 \times 5 = \square$

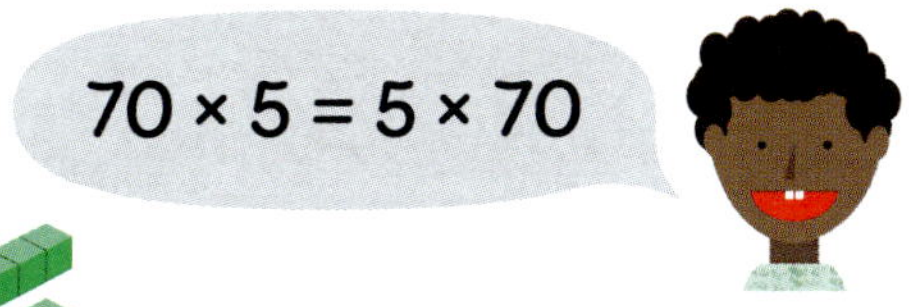

3

(a) $2 \times 30 = \square$ (b) $30 \times 3 = \square$

(c) $4 \times 50 = \square$ (d) $40 \times 5 = \square$

(e) $8 \times 30 = \square$ (f) $60 \times 8 = \square$

Complete Worksheet **1**: pages **109–110**

Multiplying 2-Digit Numbers

Lesson 2

Explore

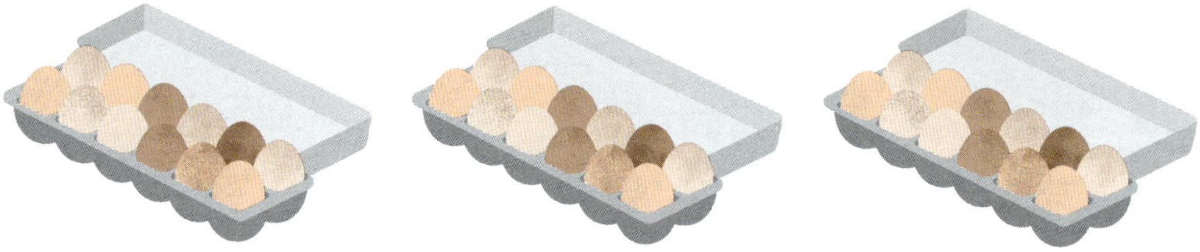

How many eggs are there in the three boxes?

Master

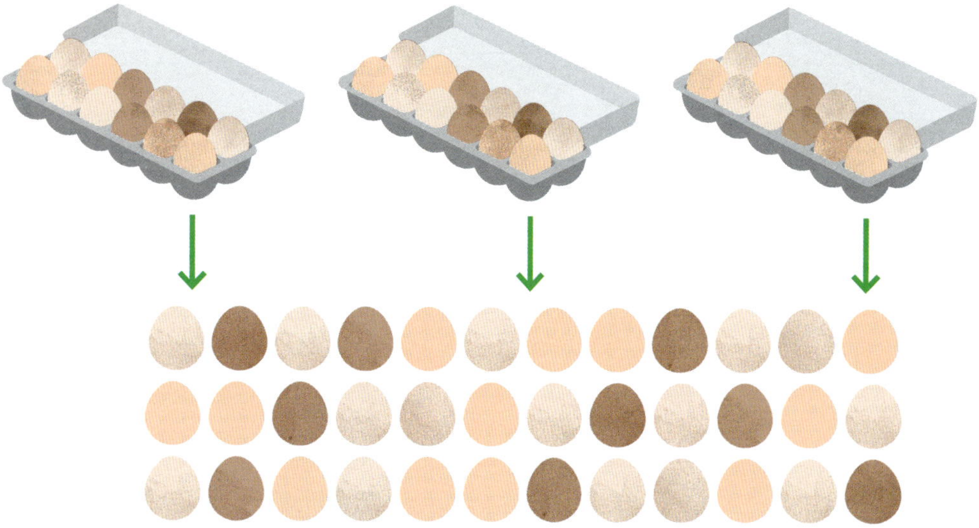

Method 1

12 + 12 + 12 = 36

There are 36 eggs in the three boxes.

Method 2

There are 3 groups of 12 eggs.

Multiply 12 eggs by 3.

$12 \times 3 = \square$

Step 1 Multiply 2 ones by 3.

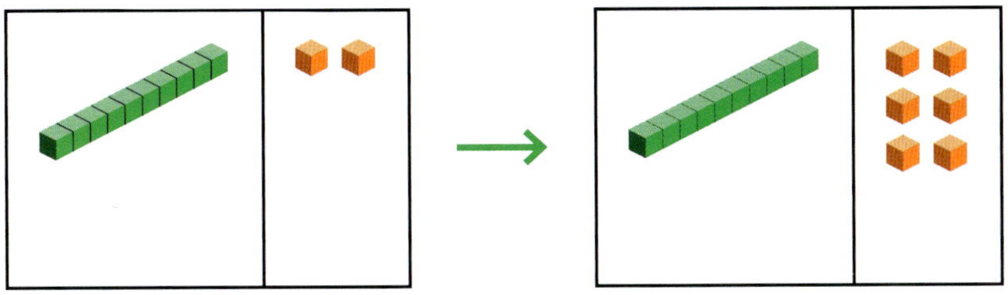

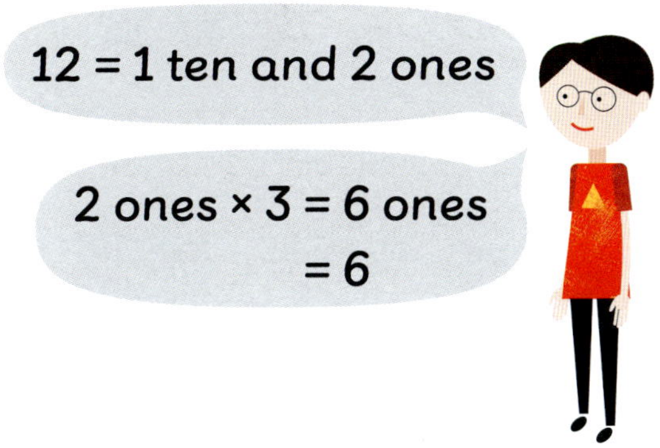

Step 2 Multiply 1 ten by 3.

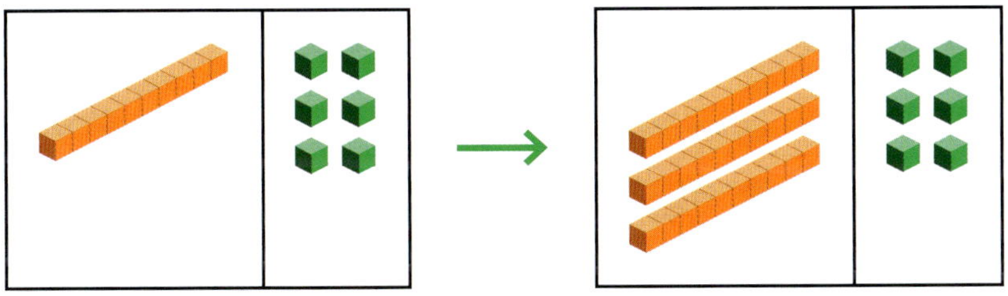

Step 3 Add the products.

$12 \times 3 = 30 + 6 = 36$

12×3

10 2

$10 \times 3 = 30$

$2 \times 3 = 6$

There are 36 eggs in the three boxes.

Guided Practice

1 Fill in the blanks.

(a)

14 × 3 = 14 + ____ + ____ = ____

(b) 14 × 3 = ____

4 × 3 = ____

10 × 3 = ____

14 × 3 = ____ + ____ = ____

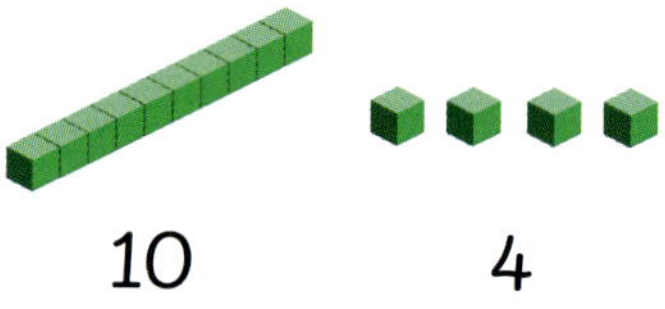

2 Multiply.

(a) 32 × 4 = ____

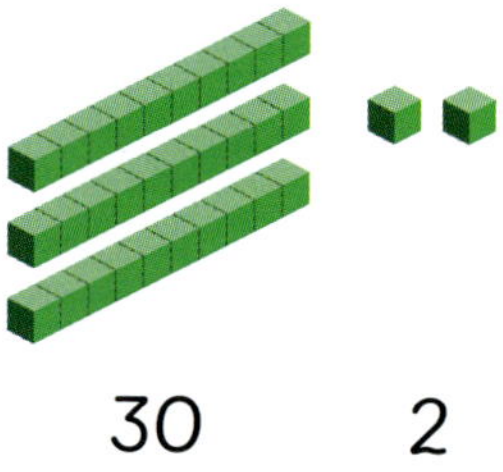

(b) 24 × 2 = ____

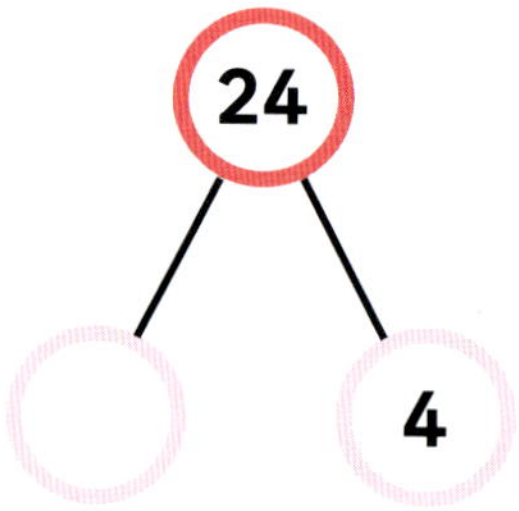

3 Multiply.

(a) 11 × 5 = ____

(b) 2 × 43 = ____

(c) 3 × 41 = ____

(d) 62 × 3 = ____

Complete Worksheet **2**: pages **111–112**

Multiplying 2-Digit Numbers

Lesson 3

Explore

There are 32 runners in each race.
How many runners are there in 3 races?

Master

32 × 3 = [ ]

Step 1 Multiply 2 ones by 3.
2 ones × 3 = 6 ones

	t	o
	3	2
×		3
		6

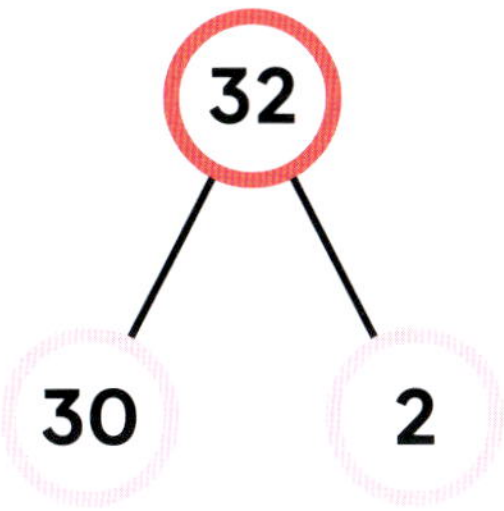

Step 2 Multiply 3 tens by 3.
3 tens × 3 = 9 tens

	t	o
	3	2
×		3
		6
	9	0

Step 3 Add the products.
6 + 90 = 96

	t	o
	3	2
×		3
		6
	9	0
	9	6

32 × 3 = 96

There are 96 runners in 3 races.

Guided Practice

1 Multiply.

(a) 13 × 2 = ____

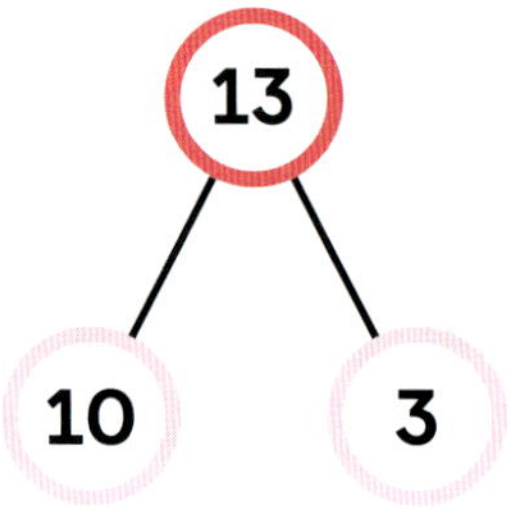

	t	o
	1	3
×		2

(b) 12 × 4 = ____

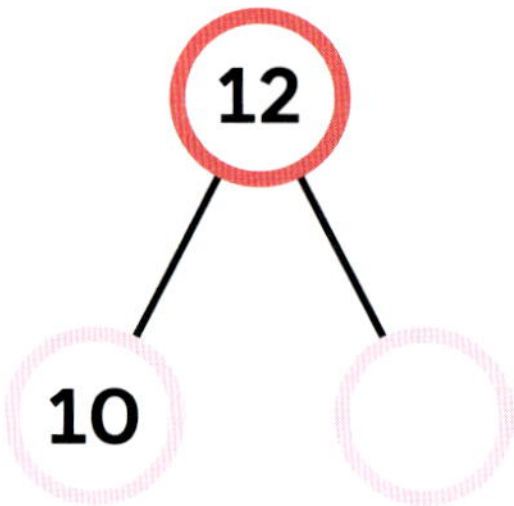

	t	o
	1	2
×		4

(c) 23 × 3 = ____

	t	o
	2	3
×		3

(d) 41 × 2 = ____

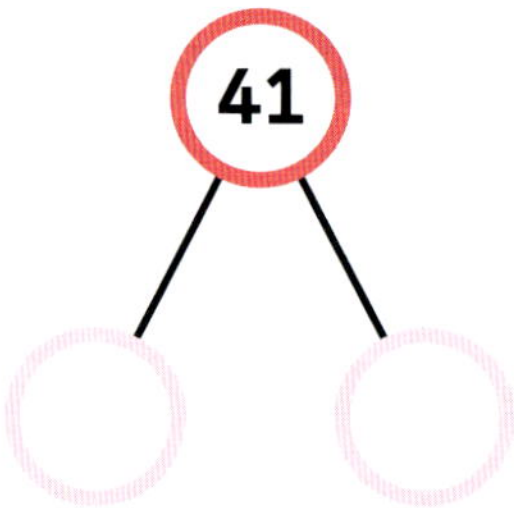

	t	o
	4	1
×		2

2 Find the product of:

(a) 14 and 2 ____

(b) 3 and 22 ____

(c) 3 and 31 ____

Complete Worksheet **3**: pages **113–114**

Multiplying with Renaming

Lesson 4

Explore

There are 26 children in each class.
How many children are in 3 classes?

Master

1 26 × 3 =

Step 1 Multiply 6 ones by 3.

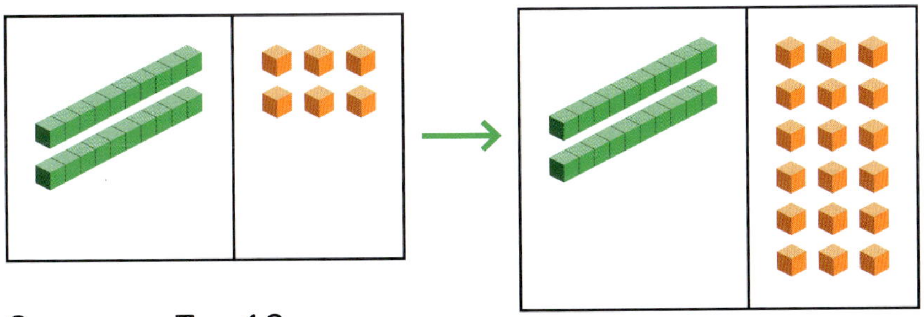

6 ones × 3 = 18 ones

Step 2 Rename the ones.

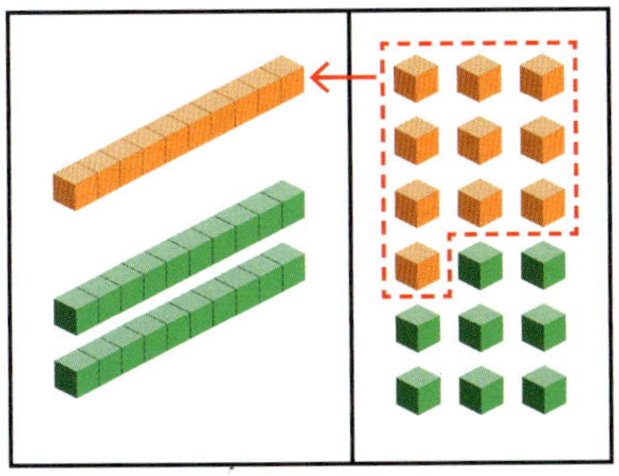

	t	o
	2	6
×		3
	1	8

18 ones = 1 ten + 8 ones

Step 3 Multiply 2 tens by 3.

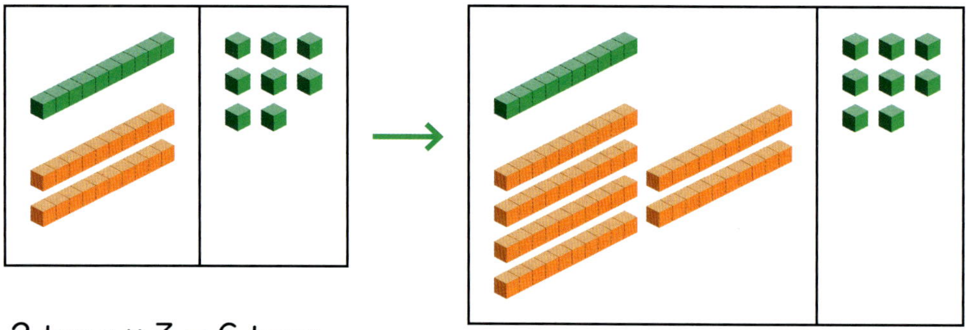

	t	o
	2	6
×		3
	1	8
	6	0

2 tens × 3 = 6 tens

Step 4 Add the products.

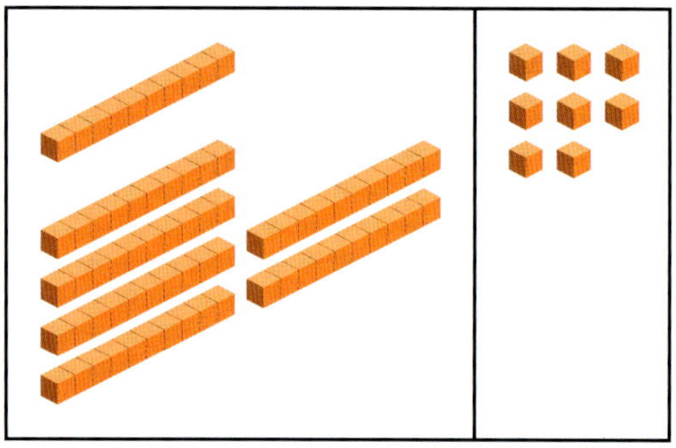

	t	o
	2	6
×		3
	1	8
+	6	0
	7	8

26 × 3 = 78

There are 78 children in 3 classes.

2 What is the product of 35 and 3?

Method 1

Step 1 Multiply 5 ones by 3.

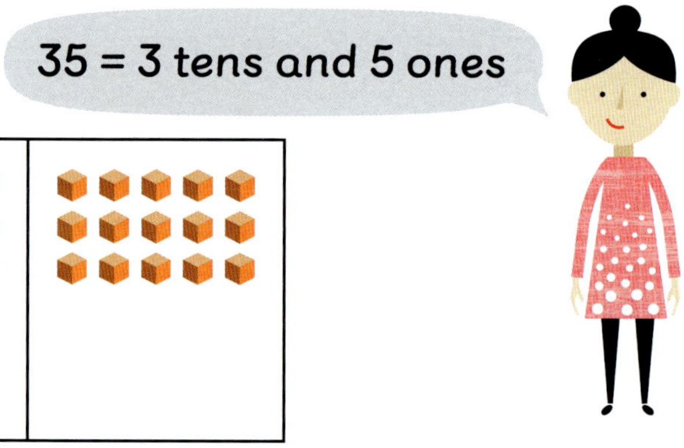

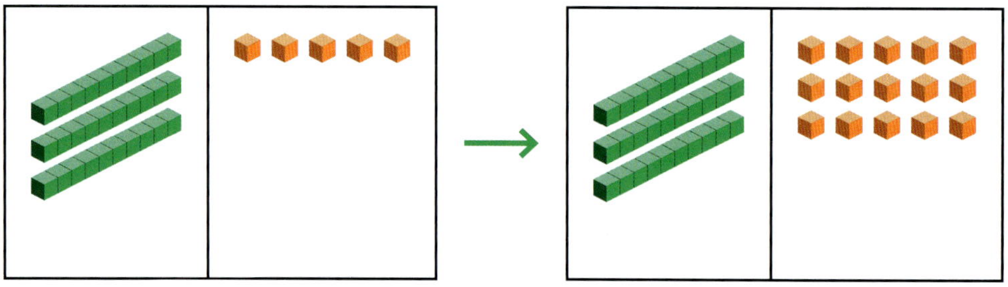

5 ones × 3 = 15 ones

Step 2 Rename the ones.

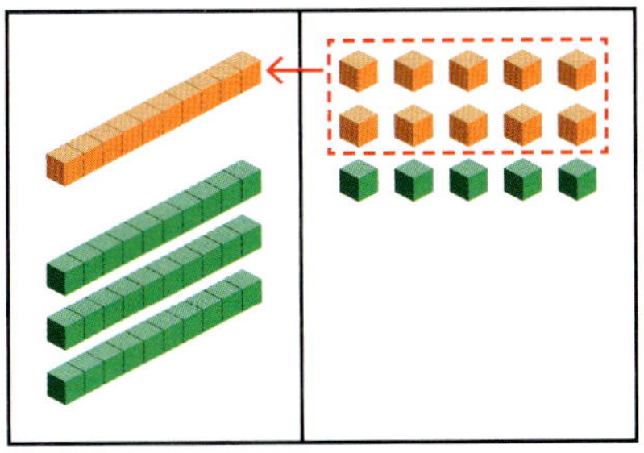

	t	o
	3	5
×		3
	1	5

15 ones = 1 ten + 5 ones

Step 3 Multiply 3 tens by 3.

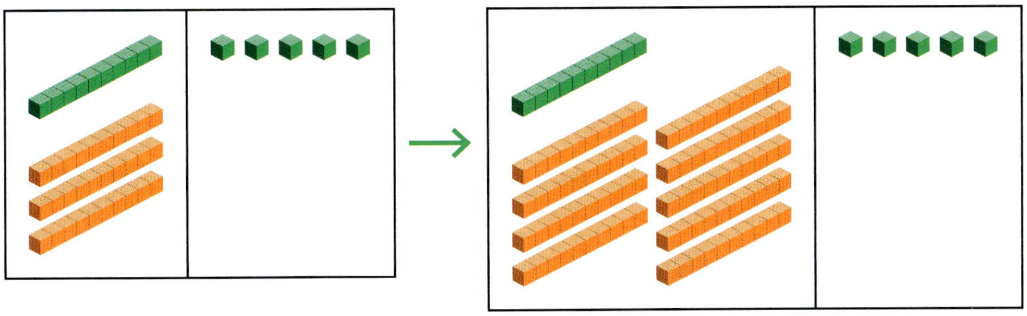

	t	o
	3	5
×		3
	1	5
	9	0

3 tens × 3 = 9 tens

Step 4 Add the products.

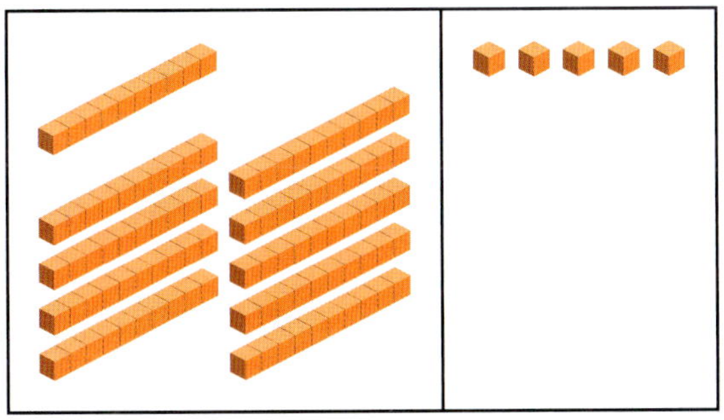

	h	t	o
		3	5
×			3
		1	5
+		9	0
	1	0	5

35 × 3 = 105

Method 2

35 × 3 is the same as 3 groups of 35.

35 × 3 = 70 + 35
= 105

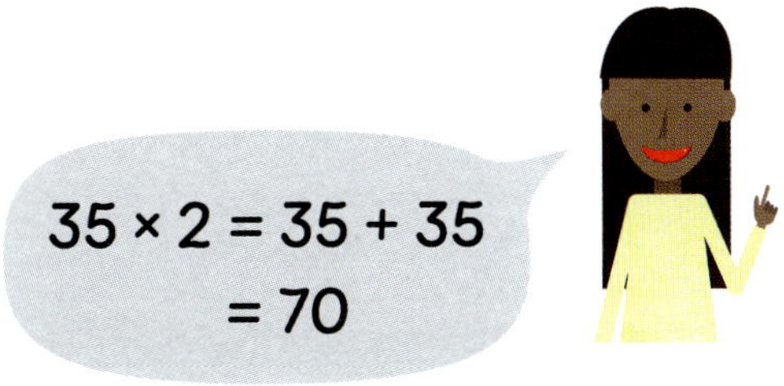

Guided Practice

1 Multiply.

(a) 16 × 4 = ☐

	t	o
	1	6
×		4
	☐	☐
+	☐	☐
	☐	☐

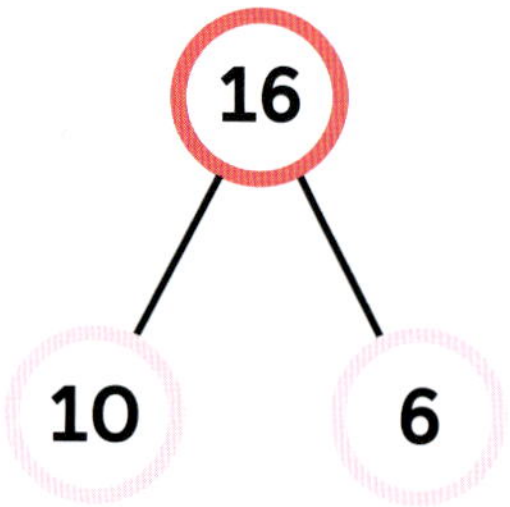

(b) 28 × 5 = ☐

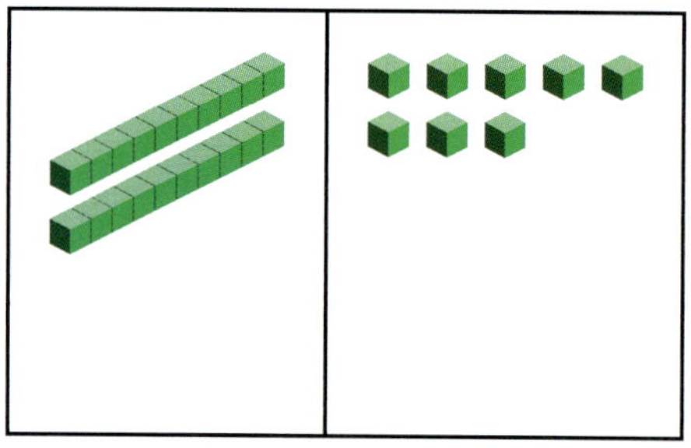

	h	t	o
		2	8
×			5
		☐	☐
+	☐	☐	☐
	☐	☐	☐

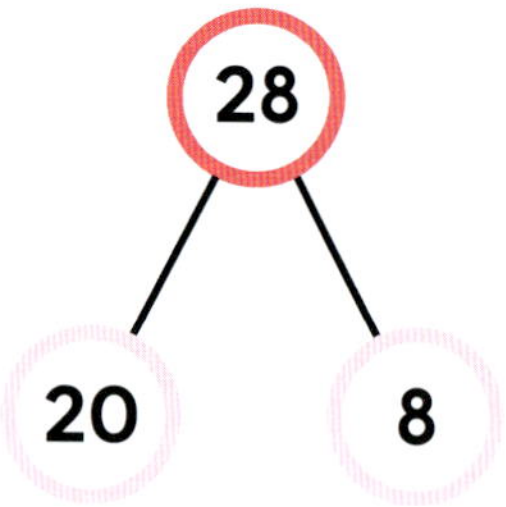

2 Find the product of:

(a) 12 and 5 ☐

(b) 27 and 4 ☐

(c) 46 and 3 ☐

(d) 8 and 38 ☐

Complete Worksheet **4**: pages **115–116**

Multiplying with Renaming

Lesson 5

Explore

Jacob has 4 boxes of mangoes. Each box contains 36 mangoes. He finds the total number of mangoes he has using the method shown. Is he correct?

h	t	o
	$^{2}3$	6
×		4
1	4	4

Master

1 36 × 4 = ☐

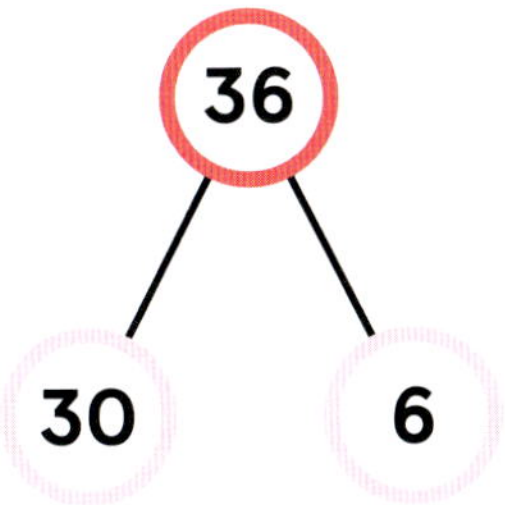

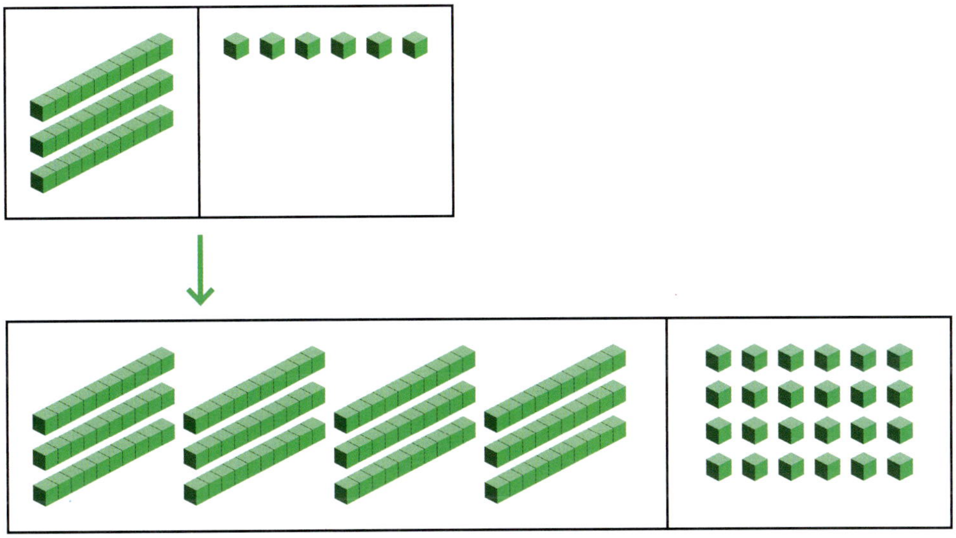

Step 1 Multiply the ones.

6 ones × 4 = 24 ones

24 ones = 2 tens + 4 ones

	t	o	
2 tens	$^{2}3$	6	
×		4	
		4	4 ones

Step 2 Multiply the tens.

3 tens × 4 = 12 tens

12 tens + 2 tens = 14 tens

h	t	o
	$^{2}3$	6
×		4
1	4	4

36 × 4 = 144

Jacob is correct.

14 tens = 1 hundred + 4 tens

2 Find the product of 43 and 8.

Step 1

	t	o
	$\overset{2}{4}$	3
×		8
		4

3 ones × 8 = 24 ones
24 ones = 2 tens + 4 ones

43 × 8 = 344

The product of 43 and 8 is 344.

Step 2

	h	t	o
		$\overset{2}{4}$	3
×			8
	3	4	4

4 tens × 8 = 32 tens
32 tens + 2 tens = 34 tens

34 tens = 3 hundreds + 4 tens

Guided Practice

1 Multiply.

(a) 48 × 2 =

	t	o
	4	8
×		2

(b) 3 × 28 =

	t	o
	2	8
×		3

2 Multiply.

(a) 5 × 29 =

(b) 32 × 8 =

3 Find the product of:

(a) 26 and 4

(b) 34 and 8

Complete Worksheet **5**: pages **117–118**

Dividing 2-Digit Numbers

Lesson 6

Explore

Holly and Emma share 48 berries equally between them.
How many berries will they each get?

Master

Divide 48 by 2 to find the number of berries they each get.

$48 \div 2 = $ ▭

Step 1 Divide 4 tens by 2.

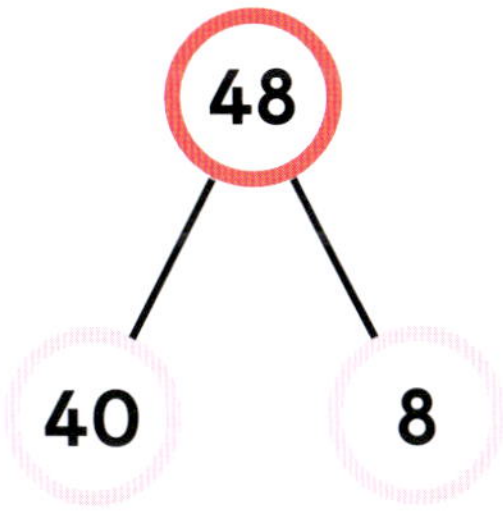

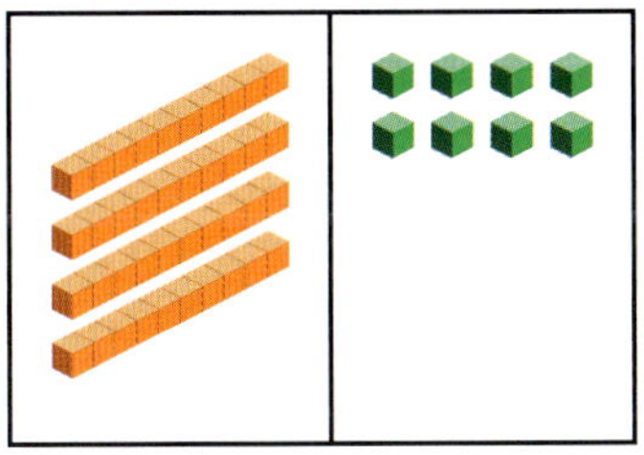

4 tens ÷ 2 = 2 tens

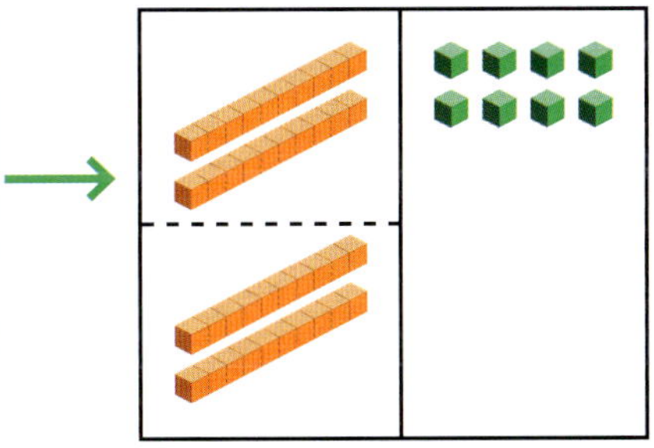

Step 2 Divide 8 ones by 2.

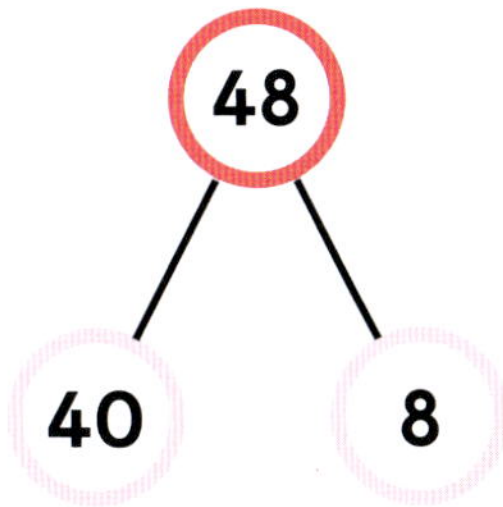

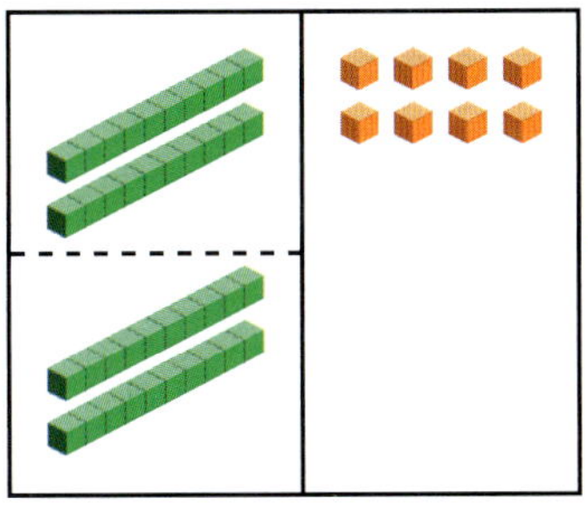

8 ones ÷ 2 = 4 ones

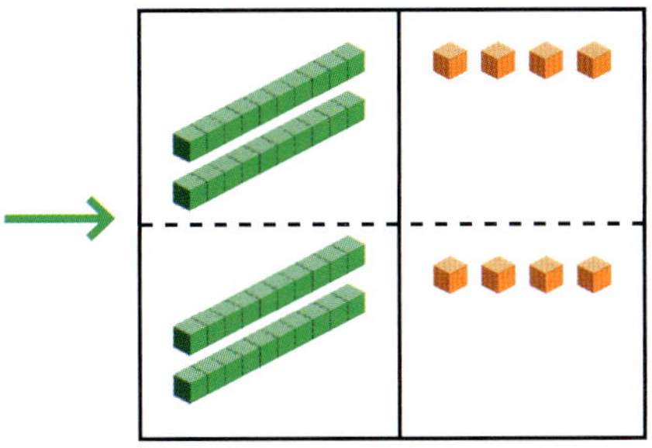

Step 3 Add the results.

$48 \div 2 = 20 + 4 = 24$

Holly and Emma get 24 berries each.

Guided Practice

Divide.

1 84 ÷ 2 = ____

84 ÷ 4 = ____

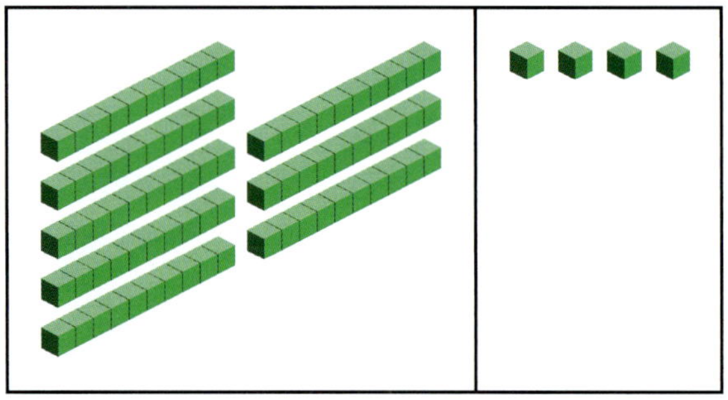

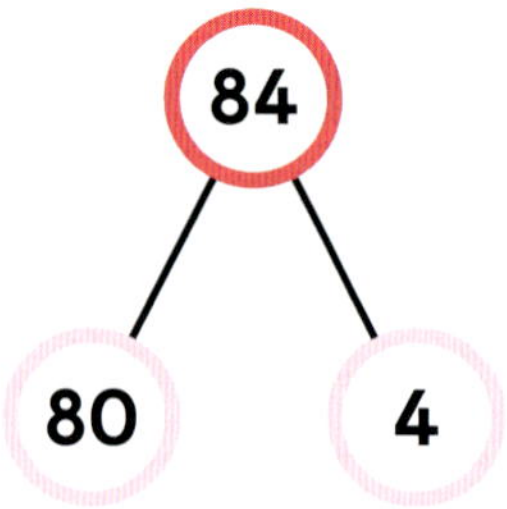

2 96 ÷ 3 = ____

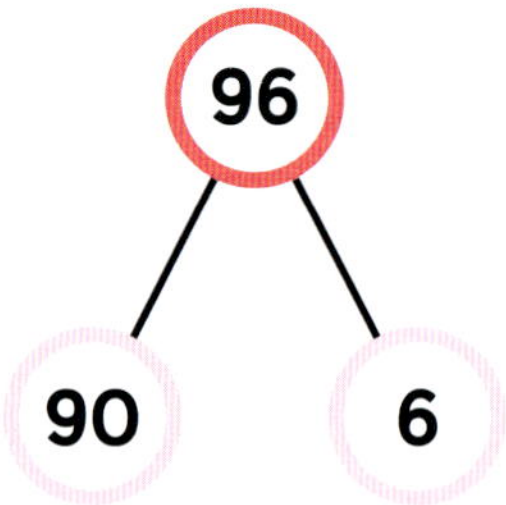

3 88 ÷ 8 = ____

4 36 ÷ 3 = ____

5 64 ÷ 2 = ____

Complete Worksheet **6**: pages **119–120**

Dividing with Renaming

Lesson 7

Explore

48 children are having music lessons in 3 classrooms.
There are an equal number of children in each classroom.

How many children are there in each classroom?

Master

Find the number of children in each classroom by dividing 48 by 3.

$48 \div 3 =$ ☐

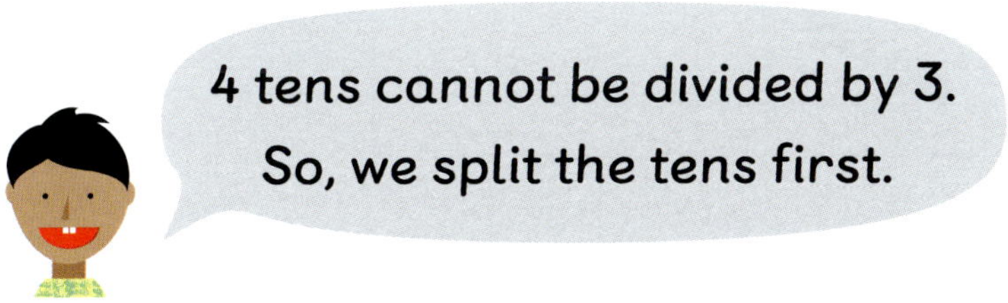

Step 1 Split 48 into 30 and 18.

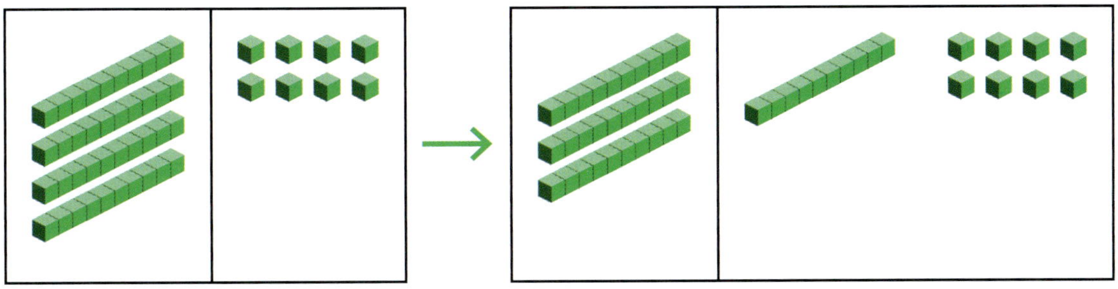

Step 2 Divide 3 tens by 3.

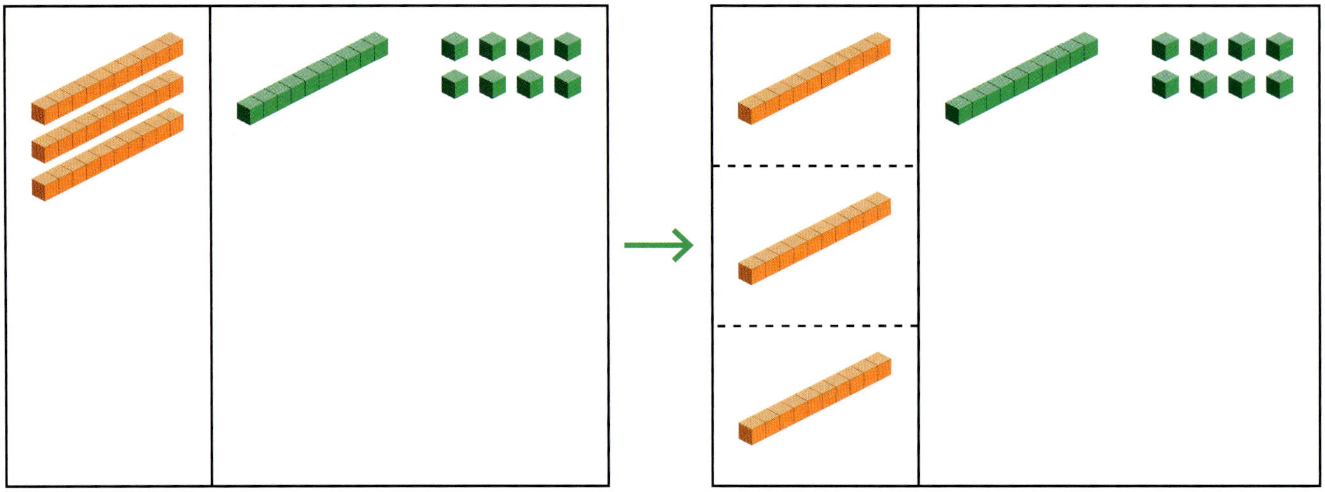

3 tens ÷ 3 = 1 ten

Step 3 Rename 1 ten as 10 ones.

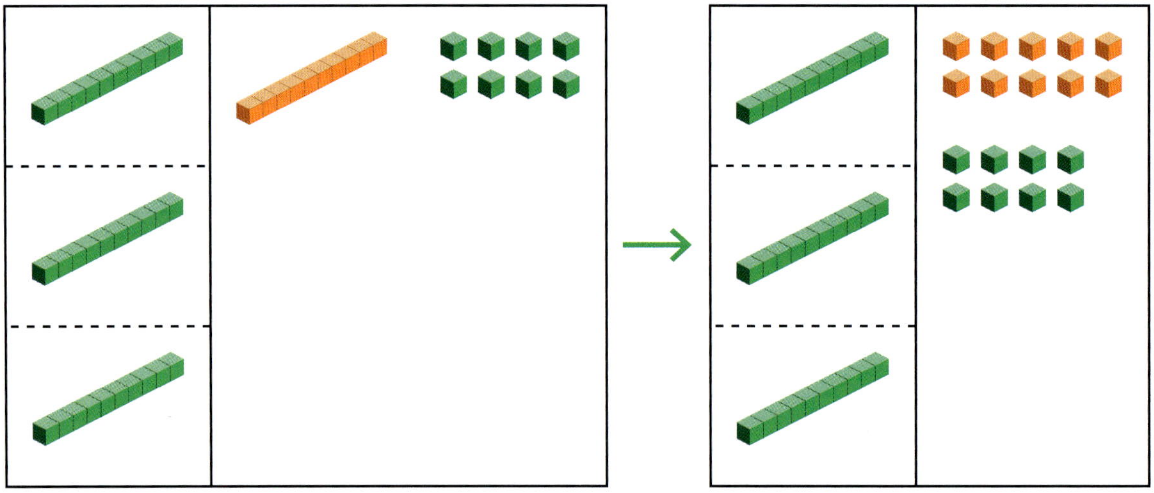

Step 4 Divide 18 ones by 3.

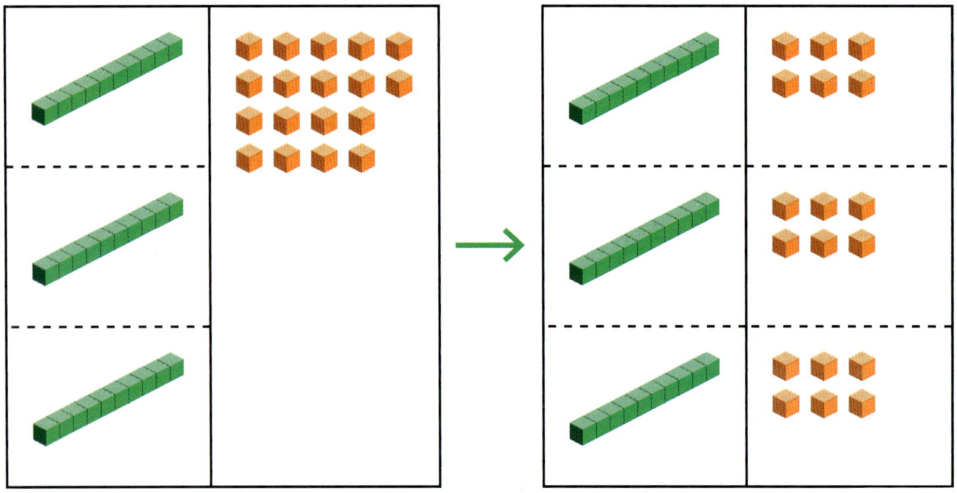

18 ones ÷ 3 = 6 ones

Step 5 Add the results.
$48 \div 3 = 1$ ten + 6 ones = 16

There are 16 children in each classroom.

When 48 is divided by 3, the quotient is 16.

When a number is divided by another number, the answer is called the **quotient**.

Guided Practice

Divide.

Use 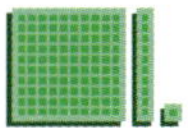 to help you.

1 $56 \div 4 =$ ▭

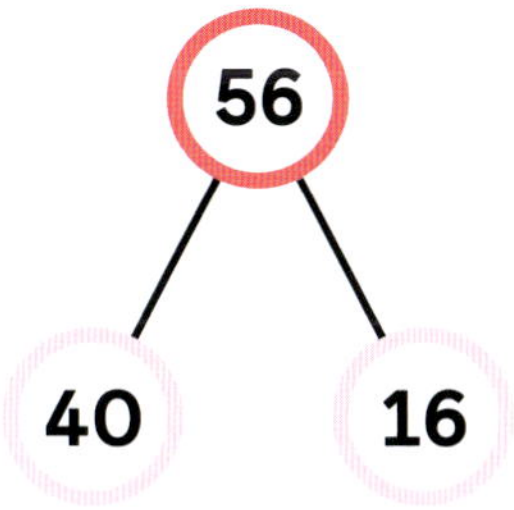

2 $96 \div 8 =$ ▭

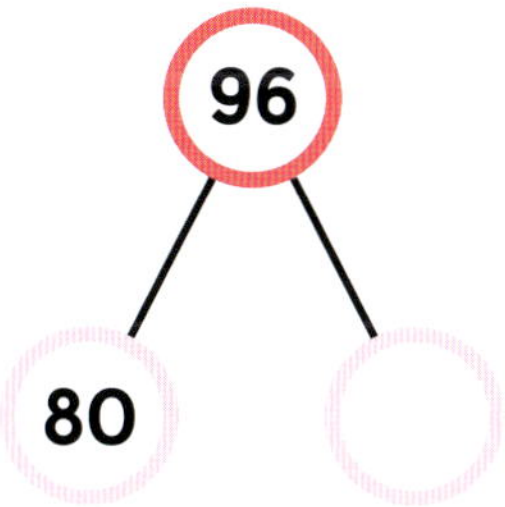

3 $78 \div 3 =$ ▭

4 $54 \div 2 =$ ▭

Complete Worksheet **7**: pages **121–122**

Dividing with Renaming

Lesson 8

Explore

Holly divided 64 by 4 using the method shown below.

$$\begin{array}{r} 16 \\ 4\overline{)\,64} \\ -\,40 \\ \hline 24 \\ -\,24 \\ \hline 0 \end{array}$$

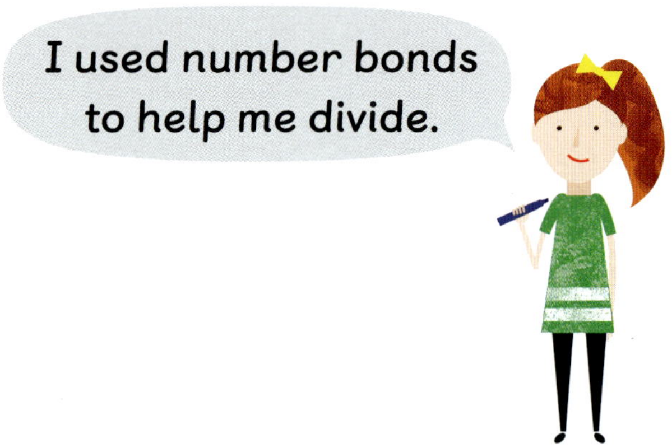

What do you think Holly means?

Master

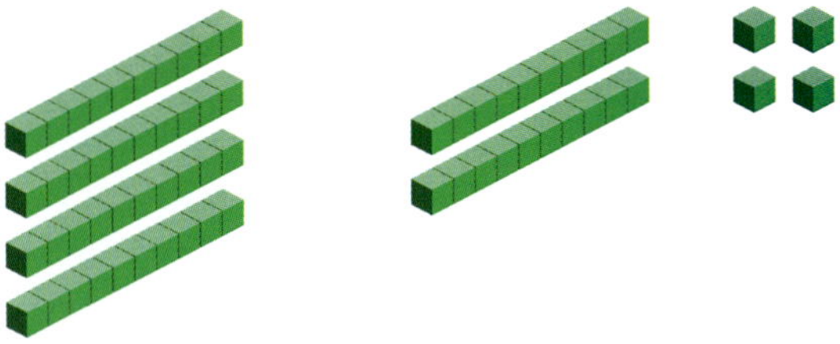

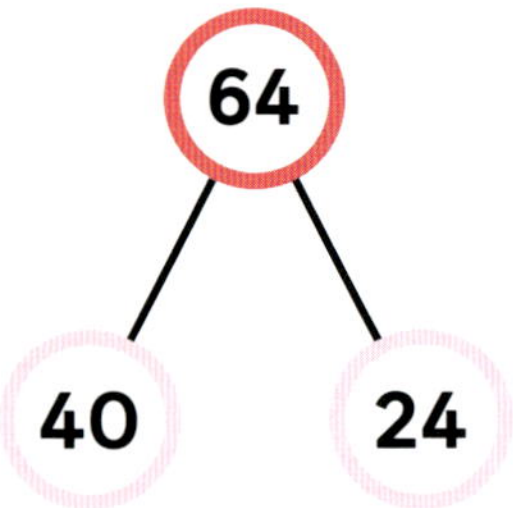

Step 1 Take 40 from 64. Divide it by 4.

$$\begin{array}{r} 1 \\ 4\overline{)\,64} \\ -\,40 \\ \hline 24 \end{array}$$

1 ten

4 tens ÷ 4 = 1 ten

Step 2 Take the remaining 24. Divide it by 4.

$$\begin{array}{r|rr} & 1 & 6 \\ \hline 4 & 6 & 4 \\ - & 4 & 0 \\ \hline & 2 & 4 \\ - & 2 & 4 \\ \hline & & 0 \end{array}$$

6 ones

24 ones ÷ 4 = 6 ones

64 ÷ 4 = 1 ten + 6 ones
= 16

Guided Practice

Divide.

1 86 ÷ 2 = ___

$$\begin{array}{r|rr} & \square & \square \\ \hline 2 & 8 & 6 \\ - & \square & \square \\ \hline & \square & \square \\ - & \square & \square \\ \hline & \square & \square \end{array}$$

2 72 ÷ 3 = ___

$$\begin{array}{r|rr} & \square & \square \\ \hline 3 & 7 & 2 \\ - & \square & \square \\ \hline & \square & \square \\ - & \square & \square \\ \hline & \square & \square \end{array}$$

3 95 ÷ 5 = ___

4 92 ÷ 4 = ___

Complete Worksheet **8**: pages **123–124**

Solving Word Problems

Lesson 9

Explore

There are 17 cars parked in the car park.
There are 2 times as many cars parked on the road.
How many cars are parked on the road?

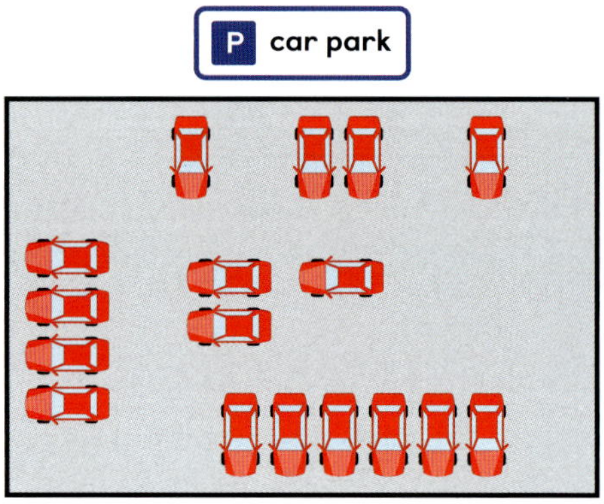

Master

1

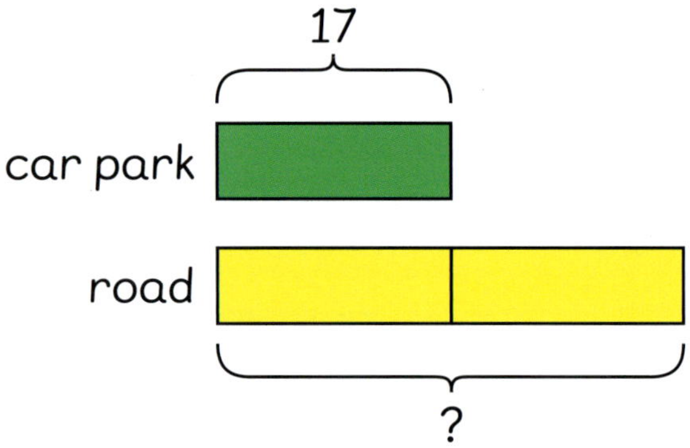

$17 \times 2 = 34$

There are 34 cars parked on the road.

2 How many cars are parked on the road and in the car park altogether?

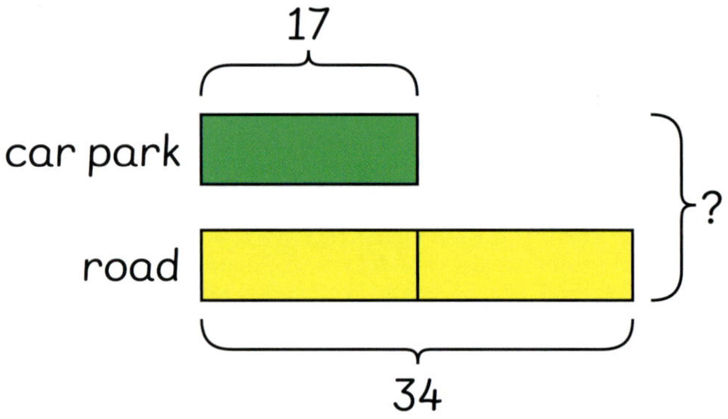

$17 + 34 = 51$

There are 51 cars parked altogether.

3 Ruby reads 26 pages of her book.
Hannah reads 3 times as many pages of her book as Ruby does.

(a) How many pages does Hannah read?

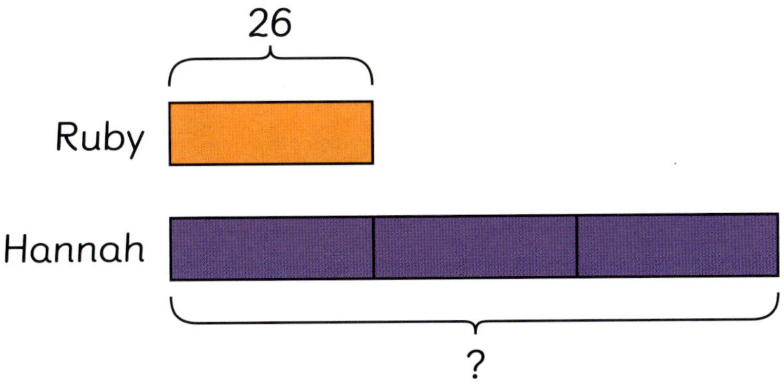

$26 \times 3 = 78$
Hannah reads 78 pages of her book.

(b) How many pages do they read altogether?

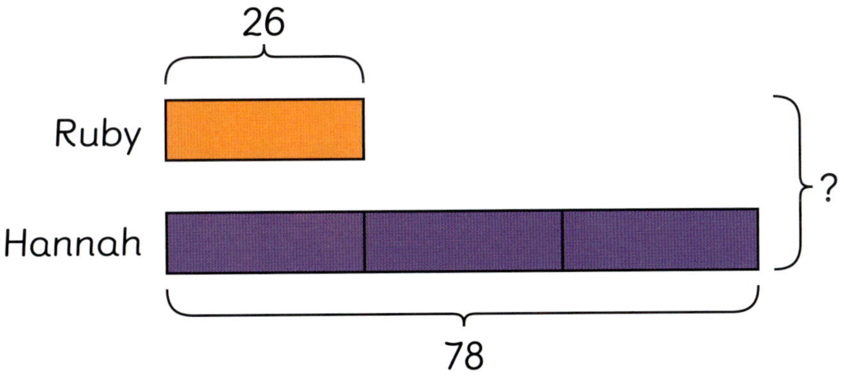

$26 + 78 = 104$
They read 104 pages altogether.

Guided Practice

Solve.

1 A farm has 27 sheep.
It has 4 times as many cows as sheep.

(a) How many cows does the farm have?

(b) How many sheep and cows does the farm have altogether?

2 There are 36 children picking some vegetables.
Each child picks 3 carrots and 3 potatoes.
How many vegetables do they pick altogether?

3 A town has 23 taxis.
On one day, each taxi takes 3 people to the airport and 5 people to the train station.
How many people take taxis in total that day?

4 Twice as many children went to the school library in the morning than in the afternoon.
14 children went to the library in the afternoon.
How many children went to the library altogether that day?

Complete Worksheet **9**: pages **125–126**

Solving Word Problems

Lesson 10

Explore

Lulu swam 62 lengths of the local swimming pool.
She swam twice as many lengths as Ravi swam.
How many lengths did the children swim altogether?

Master

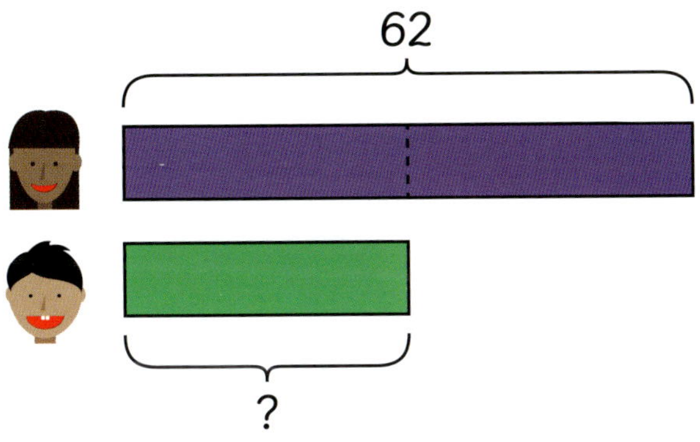

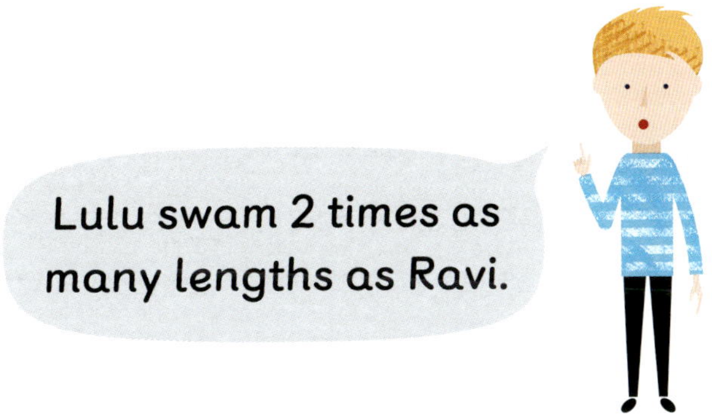

$62 \div 2 = 31$

Ravi swam 31 lengths.

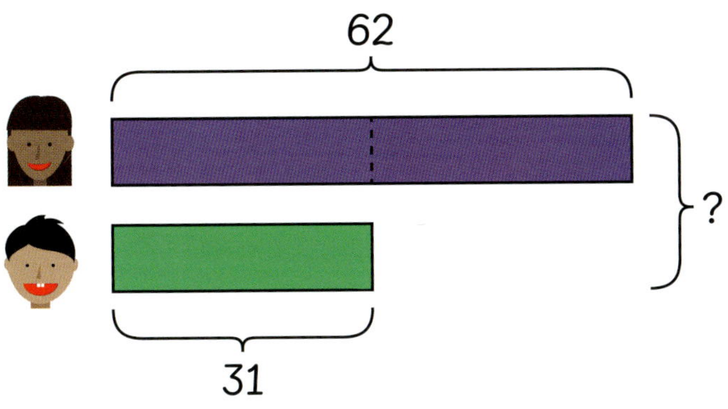

$62 + 31 = 93$

The children swam 93 lengths altogether.

Guided Practice

Solve.

1 There are three times as many green apples as there are red apples on some trees.
There are 96 green apples on the trees.

(a) How many red apples are on the trees?

(b) How many apples are on the trees altogether?

2 There are 7 times as many children travelling on a coach to the theatre as there are adults.
There are 56 people on the coach.
How many adults are on the coach?

3 In a shop, there are four times as many packets of salted crisps and cheese crisps as there are packets of popcorn.
The number of packets of salted crisps is equal to the number of packets of cheese crisps.
There are 70 packets of crisps and popcorn in the shop.
How many packets of salted crisps are there in the shop?

Complete Worksheet **10**: pages **127–128**

Solving Word Problems

Lesson 11

Explore

Ruby, Emma and Charles plant 60 seeds in total.
Ruby plants twice as many seeds as Emma plants.
Charles plants 4 more seeds than Emma plants.
How many seeds does Emma plant?

Master

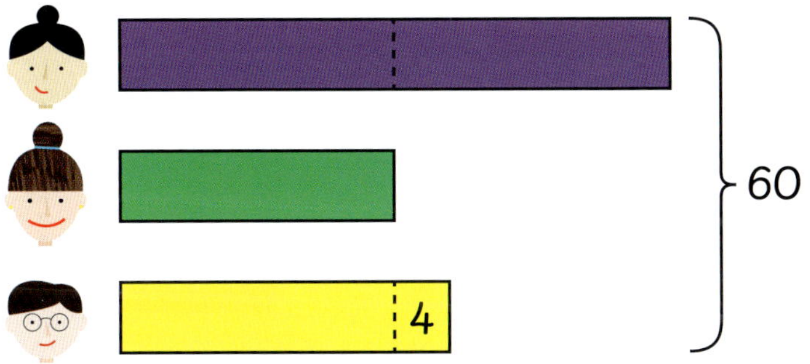

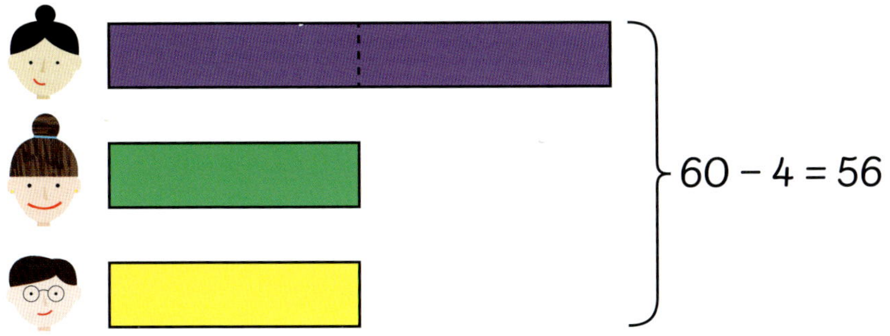

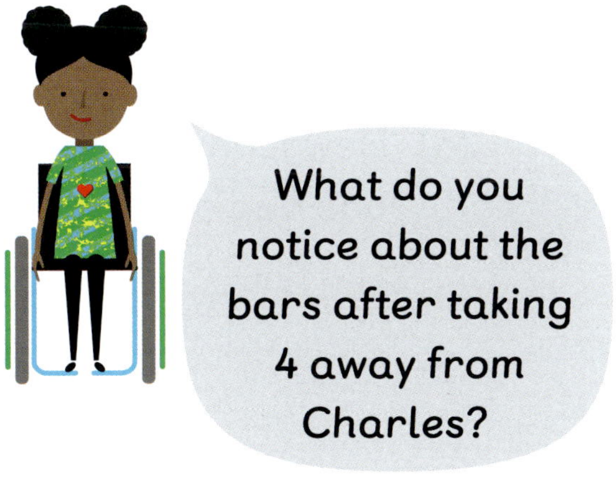

There are now 4 equal parts.

$56 \div 4 = 14$

Emma plants 14 seeds.

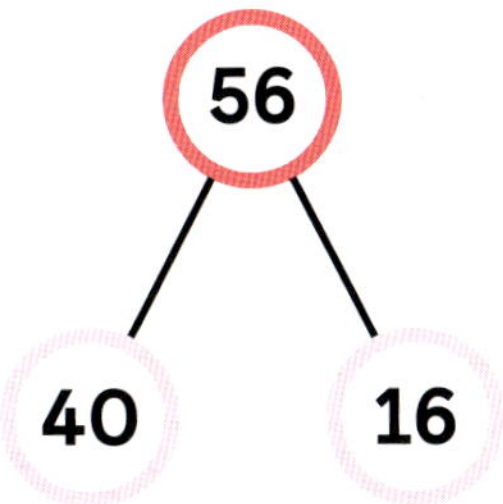

$40 \div 4 = 10$

$16 \div 4 = 4$

$56 \div 4 = 10 + 4$

$= 14$

Guided Practice

Solve.

1 87 children ride bikes and scooters to school.
15 more children ride bikes than scooters.
How many children ride scooters to school?

2 Holly, Oak and Elliott collected 35 postcards altogether.
Holly collected 6 times as many postcards as Oak collected.
Oak collected 5 more postcards than Elliott collected.
How many postcards does Oak have?

3 A restaurant makes pizza dough 4 times a day.
They use 2 kg of flour for each batch of dough.
The restaurant buys 40-kg sacks of flour.

(a) How many batches of dough can be made with one bag of flour?

(b) How many days does it take the restaurant to use an entire sack of flour?

Complete Worksheet **11**: pages **129–130**

Mind Challenge

Jacob has 8 times as many shells as Ruby has.
Amira has half as many shells as Ruby has.
If Ruby has 12 shells, how many shells do the children have in total?

My Maths Journal

Ravi started writing a word problem for his friends to solve.
Complete the problem in your journal, then ask your classmates to solve it.

Lulu, Charles and Hannah collected ______ stamps.
Lulu has ______ times as many stamps as Charles has.
Hannah has ______ (more / less) stamps than Charles has.
How many stamps does ______ have?

Review

I can...

- ☐ multiply a 2-digit number by a 1-digit number without renaming.
- ☐ multiply a 2-digit number by a 1-digit number with renaming.
- ☐ divide a 2-digit number by a 1-digit number without renaming.
- ☐ divide a 2-digit number by a 1-digit number with renaming.
- ☐ use models to solve multiplication and division word problems.

How long is Lulu's jump?

Chapter 5
Length

Writing Length in Metres and Centimetres

Lesson 1

Explore

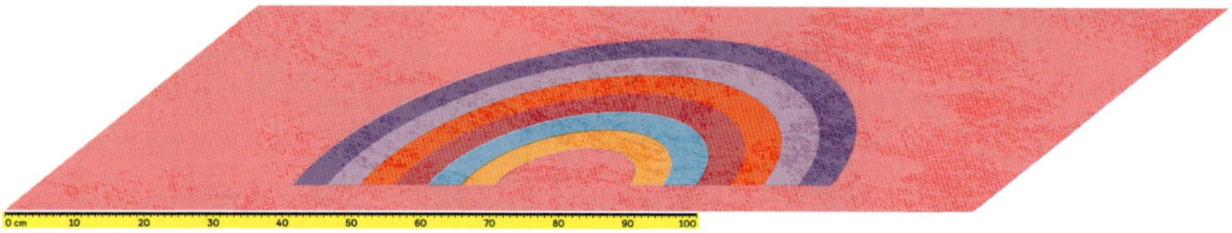

What do we know about the length of the beach towel?
How can we write its length?

Master

1 A tape measure can be used to measure lengths more than 1 metre (m).

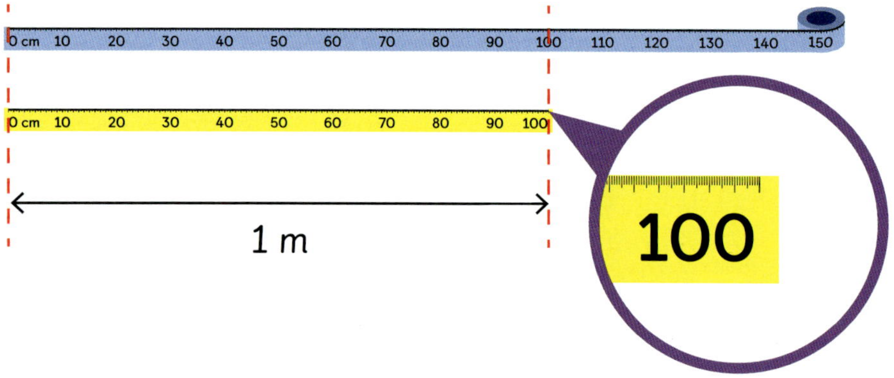

2 What is the length of the beach towel?

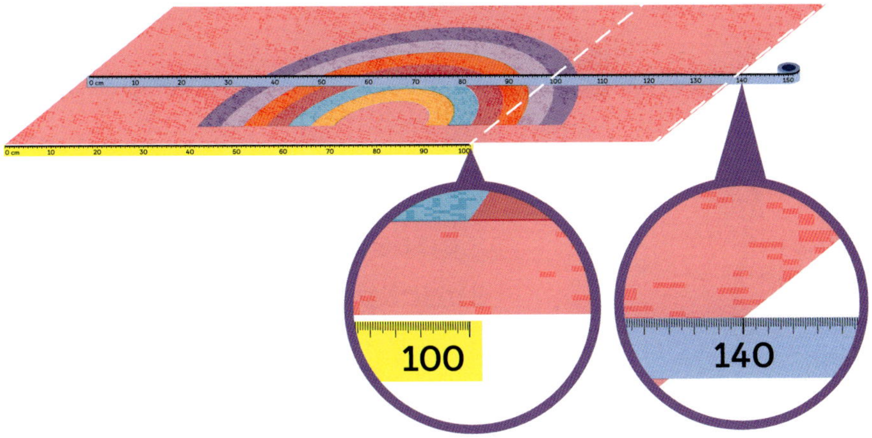

The beach towel is 140 cm long.
140 cm is longer than 1 m.

140 cm = 100 cm + 40 cm
= 1 m + 40 cm
= 1 m 40 cm

The length of the beach towel is 1 m 40 cm.

Activity Time

Group work (3 or 4 pupils)

You will need:

1. Guess the length of four objects in your classroom.
2. Measure the length of these objects using a metre ruler or tape measure.
3. Which objects are longer than 1 m? Which objects are shorter than 1 m?

Guided Practice

1 The length of the pencil case is ☐ cm.

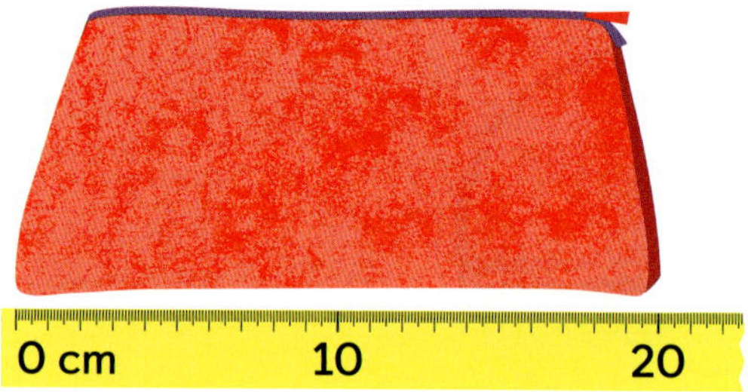

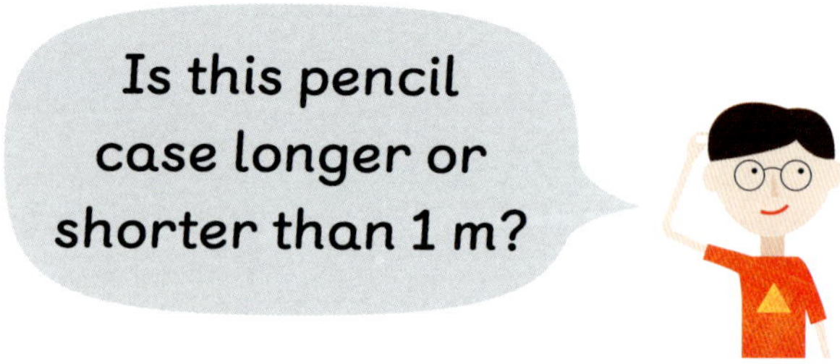

2 How wide is the bus shelter in centimetres?

The bus shelter is ☐ cm wide.

3 How tall is the house in metres and centimetres?

548 cm = ☐ m ☐ cm

548 cm

4 7 envelopes are arranged in a single row.

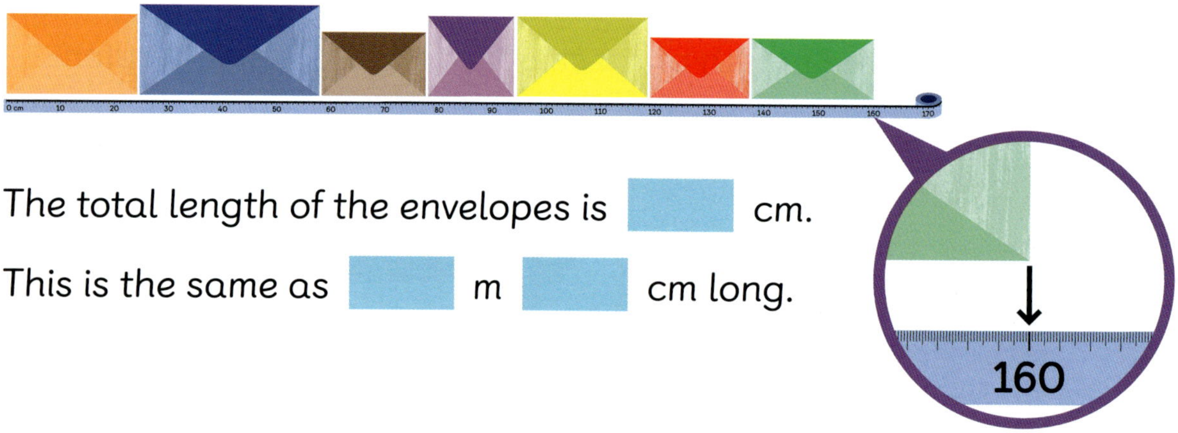

The total length of the envelopes is ☐ cm.

This is the same as ☐ m ☐ cm long.

Complete Worksheet **1**: pages **143–144**

Writing Length in Centimetres

Lesson 2

Explore

This is the length of Amira's snowboard.
What is the length of her snowboard in centimetres?

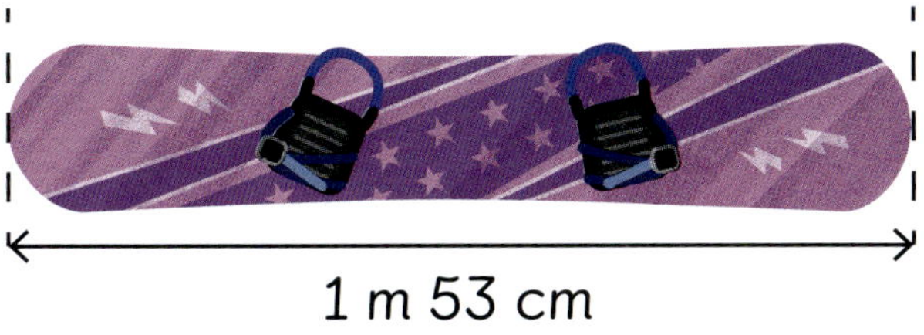

Master

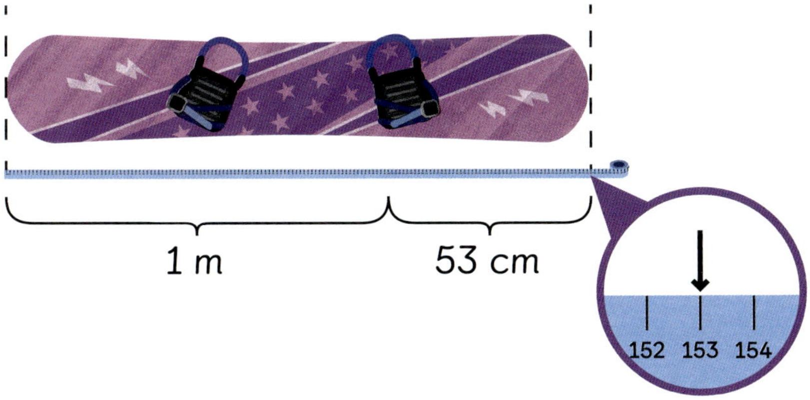

1 m 53 cm = 100 cm + 53 cm
= 153 cm

The length of Amira's snowboard is 153 cm.

Guided Practice

How long is each snowboard in centimetres?

1

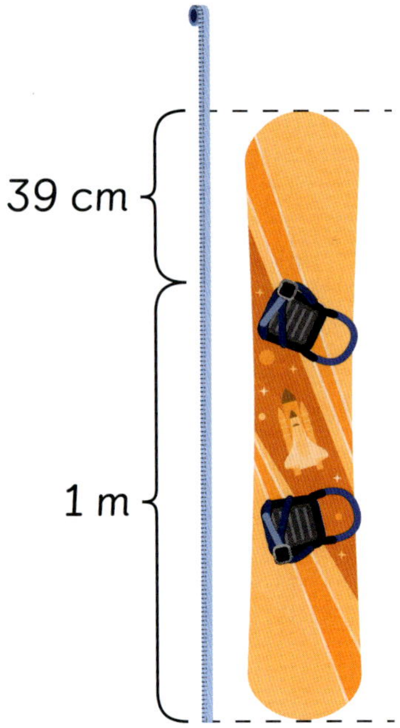

1 m 39 cm = ___ cm

2

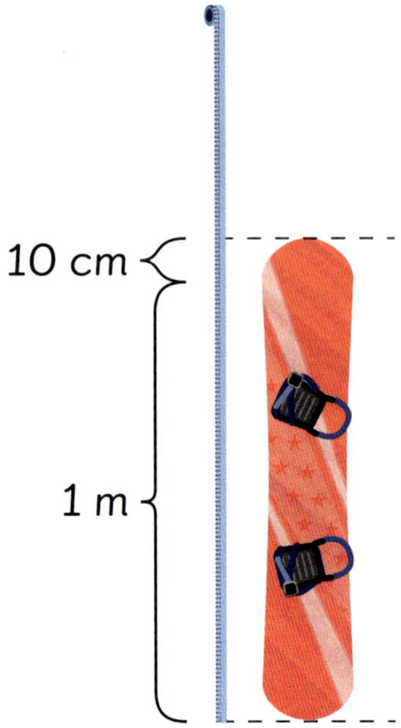

1 m 10 cm = ___ cm

3

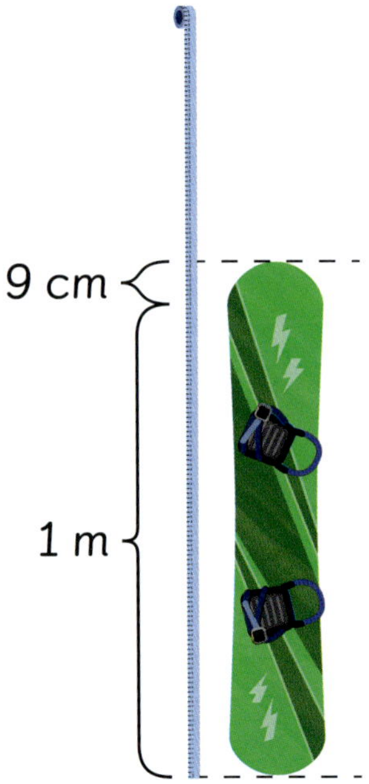

1 m 9 cm = ___ cm

4

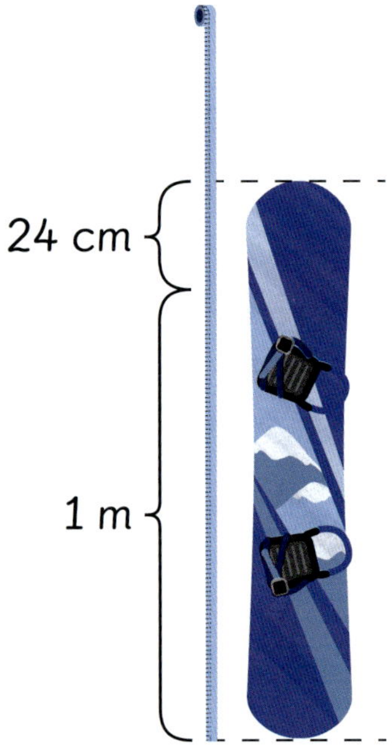

1 m 24 cm = ___ cm

Complete Worksheet **2**: pages **145–146**

Writing Length in Centimetres and Millimetres

Lesson 3

Explore

Ravi is measuring a sunflower seed.

How long is it?

Master

1

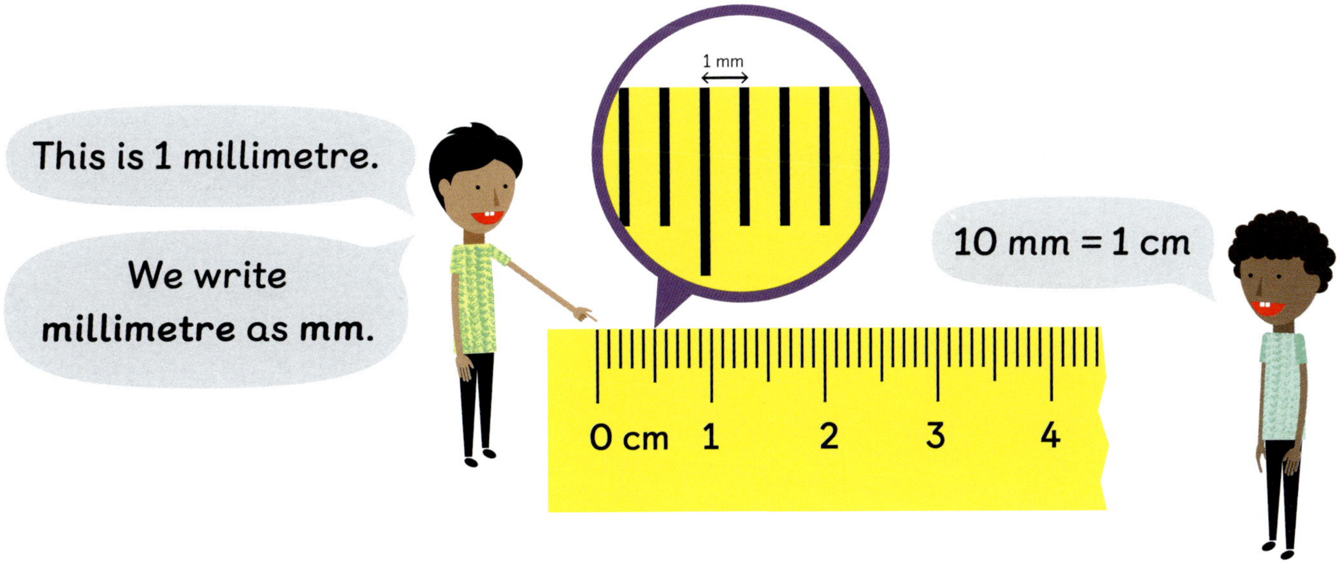

A millimetre or mm is a unit of measurement.
Millimetres or mm are used when writing the lengths of very small items.

2

The sunflower seed is 8 mm long.

3 How wide is the rubber?

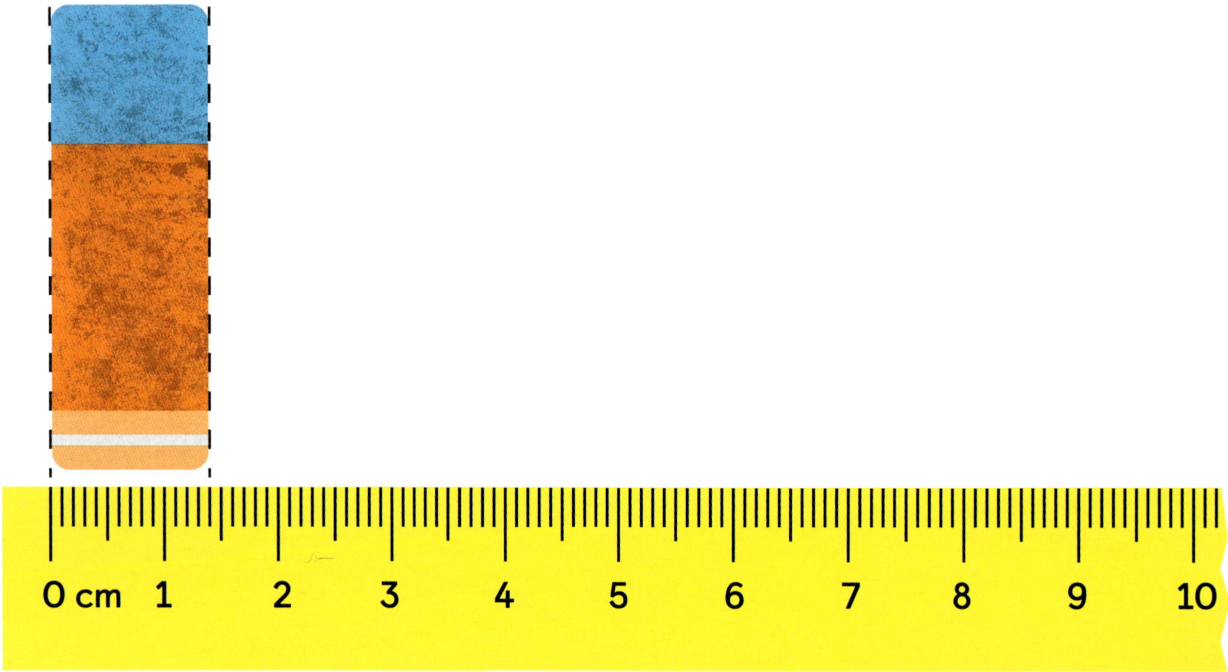

The rubber is 14 mm wide.

14 mm = 10 mm + 4 mm
= 1 cm + 4 mm
= 1 cm 4 mm

The rubber is 1 cm 4 mm wide.

4 What is 27 mm in cm and mm?

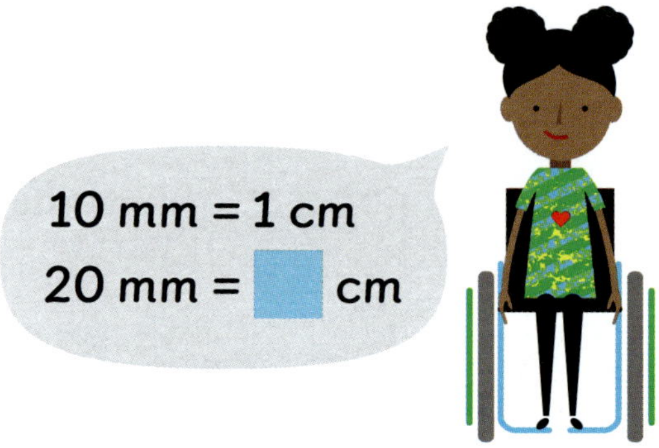

27 mm = 20 mm + 7 mm
= 2 cm + 7 mm
= 2 cm 7 mm

Guided Practice

1 Measure.

(a)

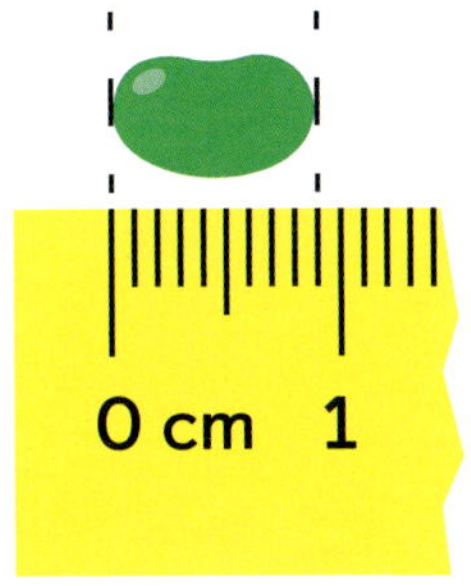

The length of the bean is ______ mm.

(b)

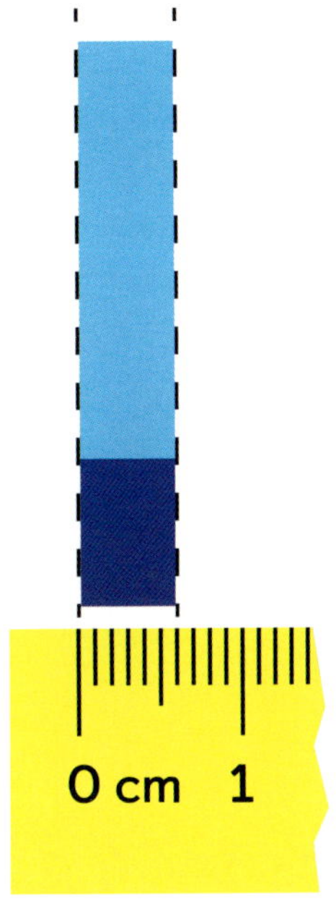

The width of the pen cap is ______ mm.

2 How long is each item in cm and mm?

(a)

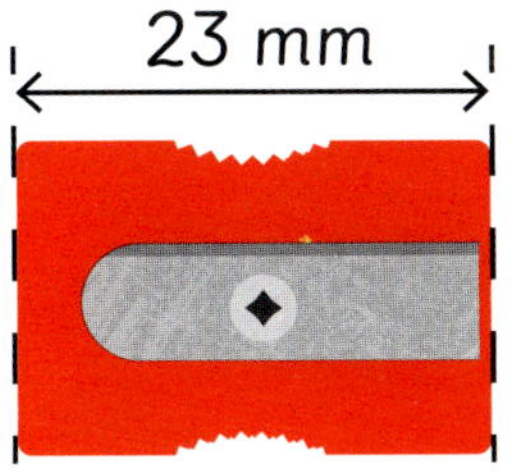

23 mm = 20 mm + ______ mm

= ______ cm + ______ mm

= ______ cm ______ mm

(b)

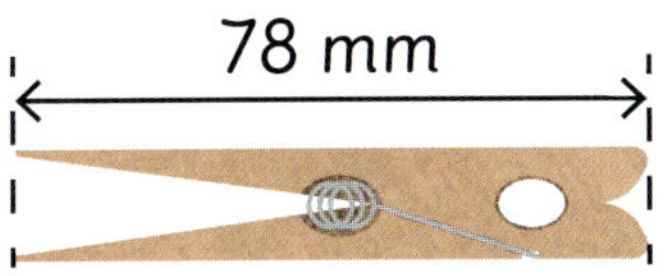

78 mm = ______ mm + ______ mm

= ______ cm + ______ mm

= ______ cm ______ mm

Complete Worksheet **3**: pages **147–148**

Writing Length in Millimetres

Lesson 4

Explore

Charles made a model car.

How long is the model car in millimetres?

Master

The model car is 7 cm 3 mm long.

7 cm 3 mm

7 cm = 70 mm   3 mm

7 cm 3 mm = 70 mm + 3 mm
= 73 mm

The model car is 73 mm long.

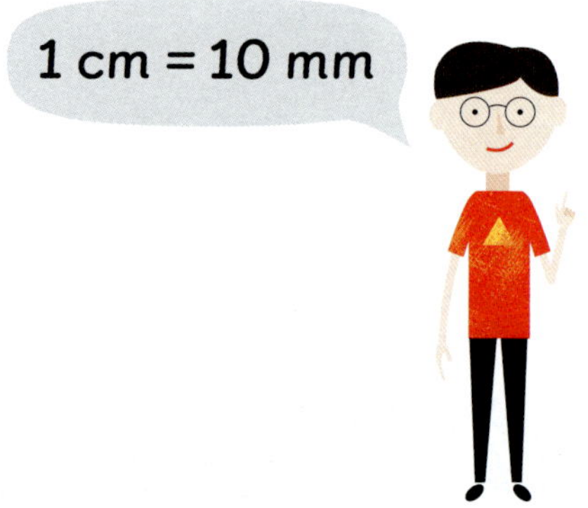

Guided Practice

1 Find the length of the crayon in millimetres.

8 cm 6 mm

8 cm = ___ mm　　6 mm

8 cm 6 mm = ___ mm + 6 mm

= ___ mm

The crayon is ___ mm long.

2 Write the lengths in millimetres.

(a) 4 cm 2 mm = ___ mm + ___ mm

= ___ mm

4 cm 2 mm

4 cm = ___ mm　　___ mm

(b) 9 cm 8 mm = ___ mm + ___ mm

= ___ mm

Complete Worksheet 4: pages **149–150**

Comparing Lengths

Lesson 5

Explore

These are the results of a long jump competition.

Lulu
135 cm

Sam
104 cm

Ruby
1 m 40 cm

Hannah
97 cm

Emma
1 m 45 cm

Who jumped further, Sam or Ruby?

Master

Method 1

104 cm = 100 cm + 4 cm
= 1 m + 4 cm
= 1 m 4 cm

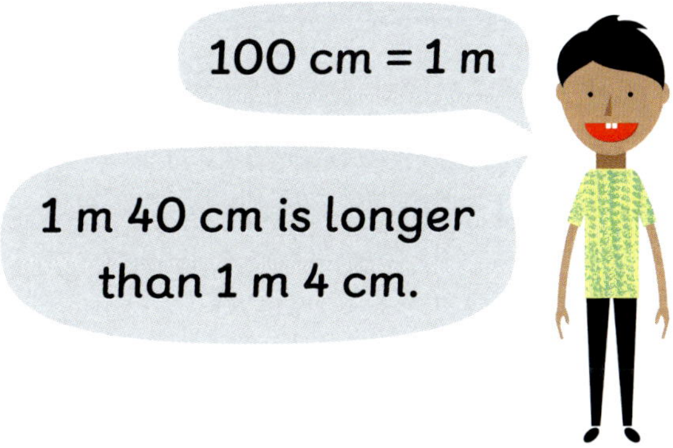

Method 2

1 m 40 cm = 100 cm + 40 cm
= 140 cm

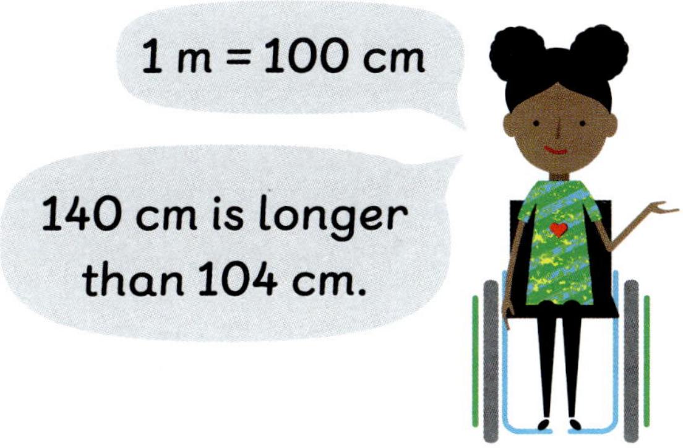

	cm	m and cm
	104 cm	1 m 4 cm
	140 cm	1 m 40 cm

Ruby jumped further than Sam.

Guided Practice

Solve the word problems.

1 Jacob has a toy boat and a toy car. The toy boat is 9 cm 7 mm long. The toy car is 99 mm long. Which is the longest toy?

2 Emma, Ruby and Amira each did the high jump at school. The table below shows the height of each child's best jump. Who jumped the highest?

Child	Height of jump
Emma	1 m 12 cm
Ruby	99 cm
Amira	105 cm

3 Wire A is 3 m 2 cm long. Wire B is 320 cm long. Wire C is 32 cm long. Arrange these wires in order from shortest to longest.

____ , ____ , ____

shortest longest

Complete Worksheet **5**: pages **151–152**

Solving Word Problems

Lesson
6

Explore

Hannah rides her bike in the park.
She rides 123 m around Section 1 and 414 m around Section 2.
How far does she ride in total?

Master

1

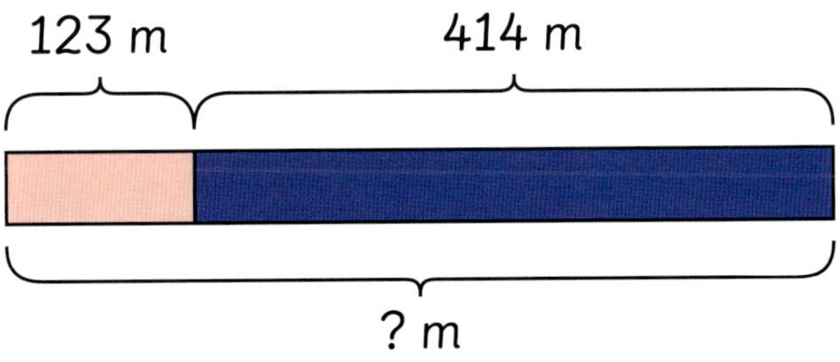

123 + 414 = 537

Hannah rides 537 m in total.

2 Sam and his mum are painting a bench that is 300 cm long. They have painted 115 cm of the bench so far.
How much of the bench is left to paint?

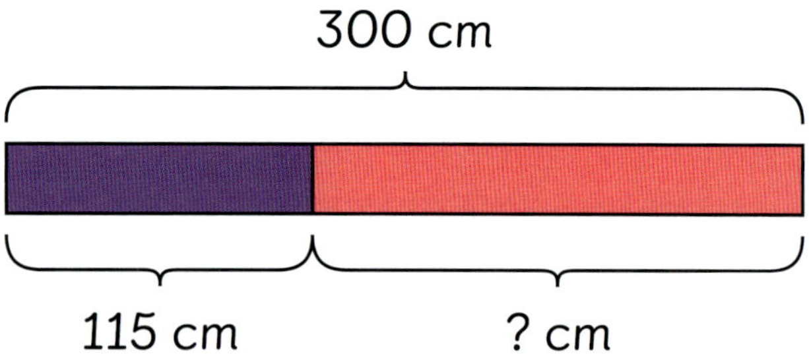

300 – 115 = 185

They still have to paint 185 cm of the bench.

Guided Practice

1

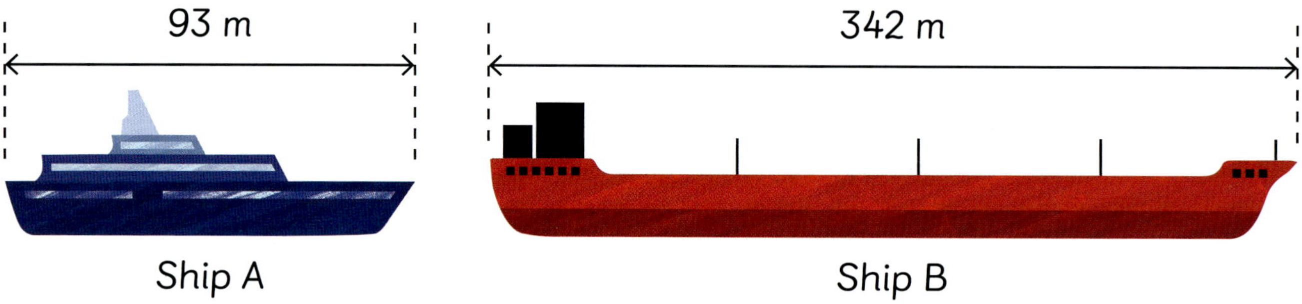

How much shorter is Ship A than Ship B?

2 Lulu sews three buttons in a row on her jacket.
What is the total width of the three buttons?

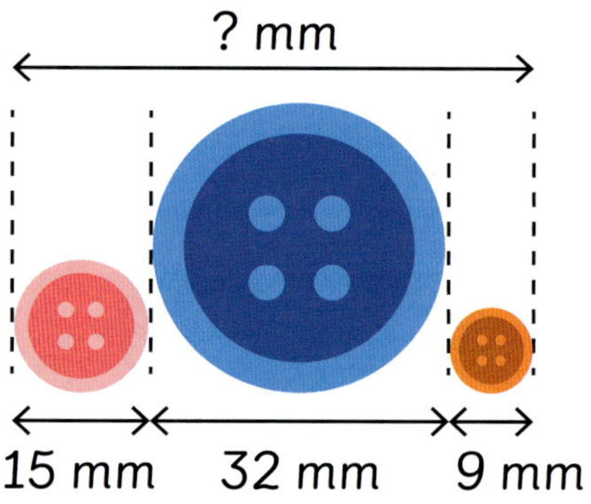

Complete Worksheet **6**: pages **153–154**

Solving Word Problems

Lesson 7

Explore

Emma had 6 m of fishing line.
She used 85 cm of it.
How much fishing line does Emma have left?

Master

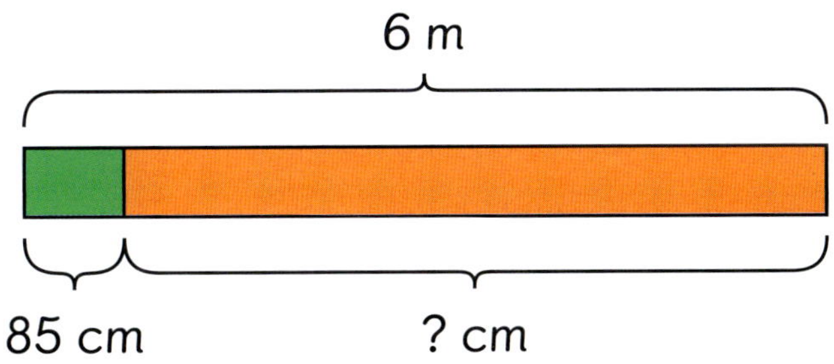

600 − 85 = 515

Emma has 515 cm of fishing line left.

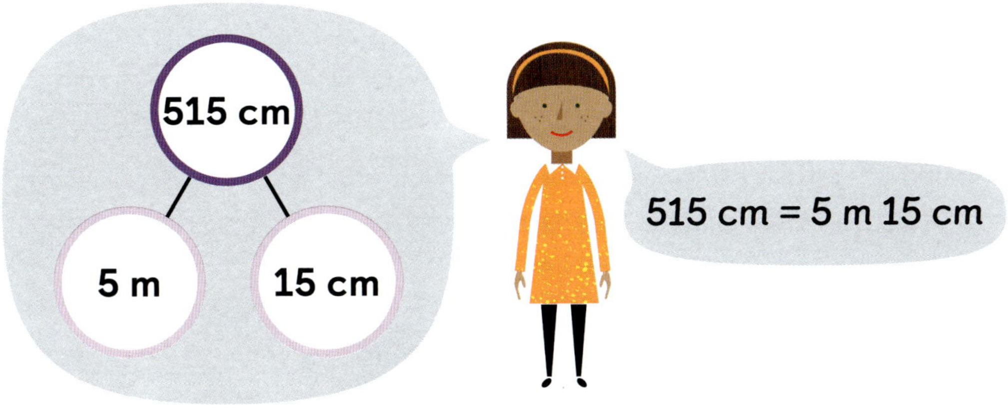

Guided Practice

1 Holly's sunflower is 78 cm taller than Ravi's sunflower.
The height of Ravi's sunflower is 1 m 39 cm.
How tall is Holly's sunflower in centimetres?

2 A baby giraffe is 1 m 89 cm tall.
A baby elephant is 97 cm tall.
What is the difference in height between the baby giraffe and the baby elephant in centimetres?

3 Charles made two towers from some blocks.
The red tower is 5 cm high. The blue tower is 34 mm high. He puts the blue tower on top of the red tower.
What is the total height of the red and blue tower in centimetres and millimetres?

4 Elliott had a 2-m wooden board.
He used some of it to build a box and had 52 cm of it left.
How much of the wooden board did he use in metres and centimetres?

5 A carton is placed next to a large box. The carton is 85 cm wide.
The box is 1 m 58 cm wide.
What is the total width of the carton and the box in centimetres?

Complete Worksheet **7**: pages **155–156**

Solving Word Problems

Lesson 8

Explore

Miss Fathima uses 4 m of yarn to knit a bag.
She makes 38 bags and has 6 m of yarn left over.
How much yarn did Miss Fathima have to start with?

Master

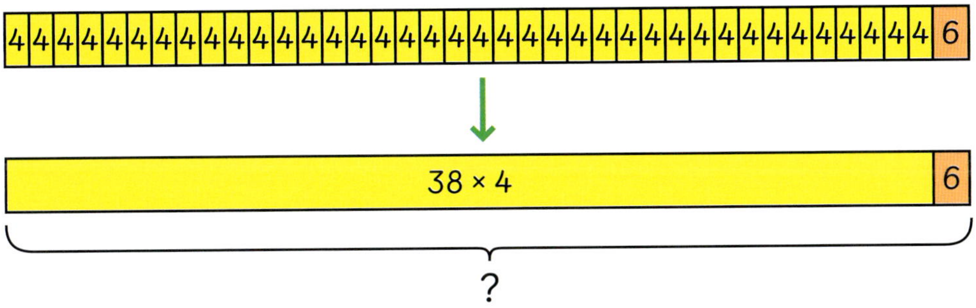

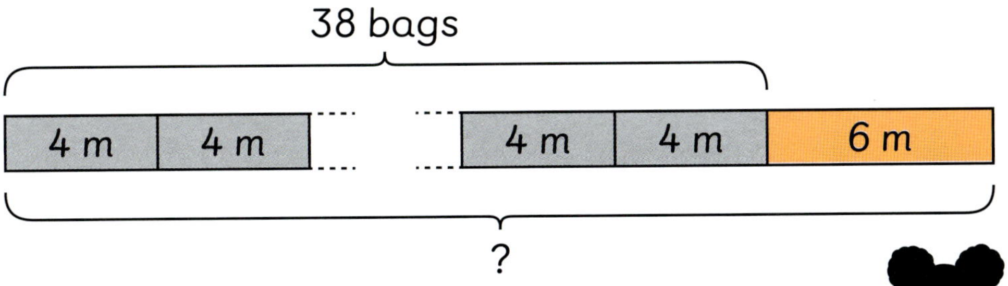

$38 \times 4 = 152$

$152 + 6 = 158$

Miss Fathima started with 158 m of yarn.

Guided Practice

1 The height of a classroom is 3 times the height of a classroom table.
The classroom table is 97 cm tall.
What is the height of the classroom?
Give your answer in metres and centimetres.

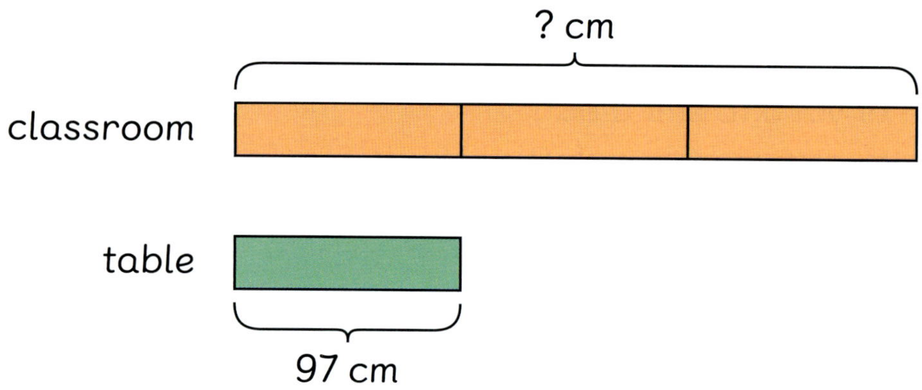

2 Holly's paper plane flew a distance of 85 cm.
Ravi's paper plane flew 5 times the distance that Holly's paper plane flew.
What is the total distance both the paper planes flew?

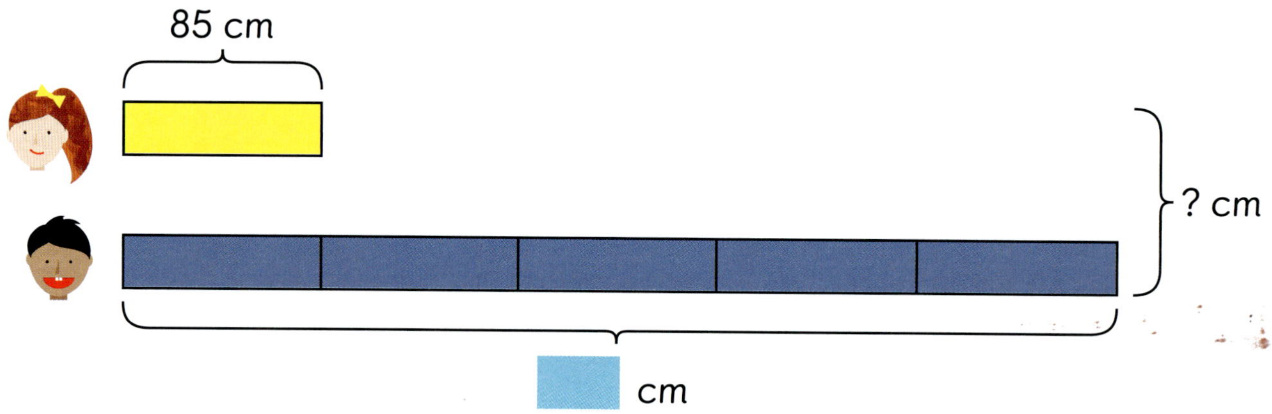

3 Charles draws a rectangle that is 35 mm long.
Sam draws a rectangle that is 4 times as long as Charles' rectangle.
What is the difference between the length of the two rectangles in centimetres and millimetres?

Complete Worksheet **8**: pages **157–158**

Solving Word Problems

Explore

The height of an apartment block is 4 times as high as the height of the car park next to it.
The apartment block is 68 m tall.
How tall is the car park?

Master

Draw a model to help you solve the problem.

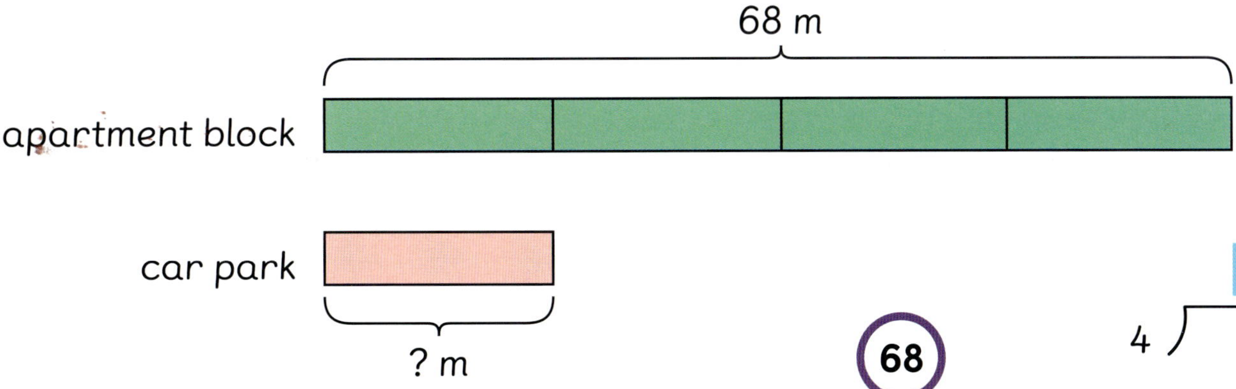

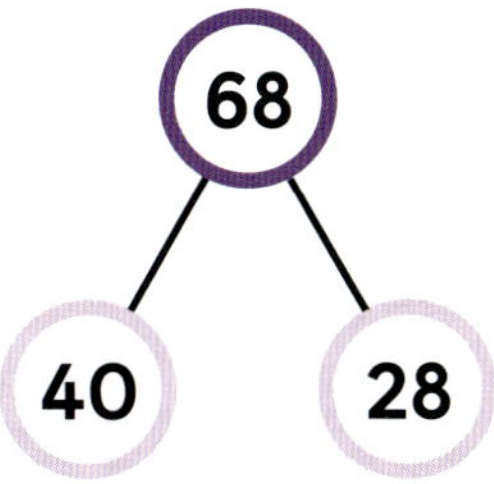

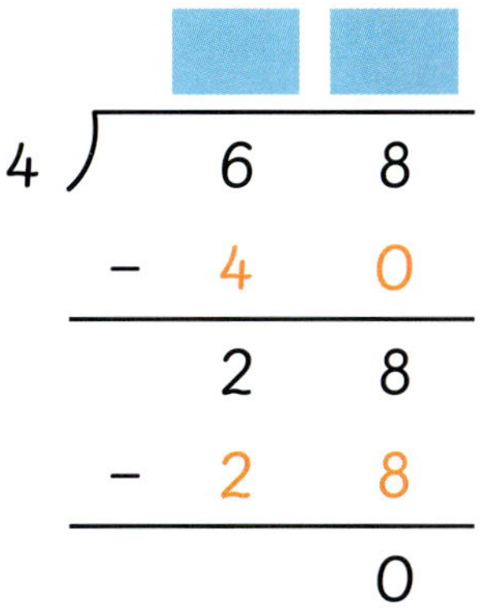

$68 \div 4 = 17$

The car park is 17 m tall.

Guided Practice

1 Lulu's mum used 4 times as much fencing in the back garden than in the front garden.
She used 35 m of fencing altogether.
How much fencing did she use in the front garden?

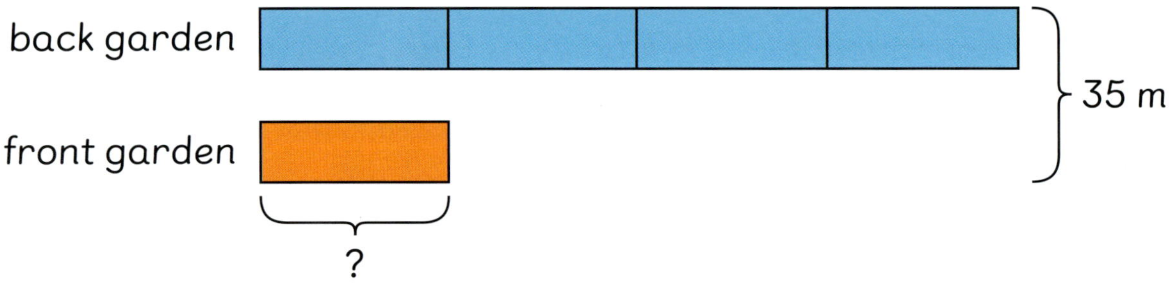

2 The yellow yo-yo is 3 times as long as the purple yo-yo.
The yellow yo-yo is twice as long as the green yo-yo.
The purple yo-yo is 26 cm long.

(a) Find the length of the yellow yo-yo.

(b) Find the length of the green yo-yo.

3 Emma, Amira and Ruby each cut a strip of paper. Emma's strip of paper is twice as long as Amira's strip of paper. Ruby's strip of paper is 3 times as long as Amira's strip of paper. The length of Ruby's strip of paper is 93 mm.

(a) What is the length of Amira's strip of paper?

(b) What is the length of Emma's strip of paper?

Complete Worksheet **9**: pages **159–160**

Solving Word Problems

Lesson 10

Explore

Holly had 140 cm of wrapping paper to wrap some presents.
Each present used the same length of wrapping paper.
After wrapping 4 presents, she had 44 cm of wrapping paper left.
What was the length of wrapping paper Holly used to wrap each present?

Master

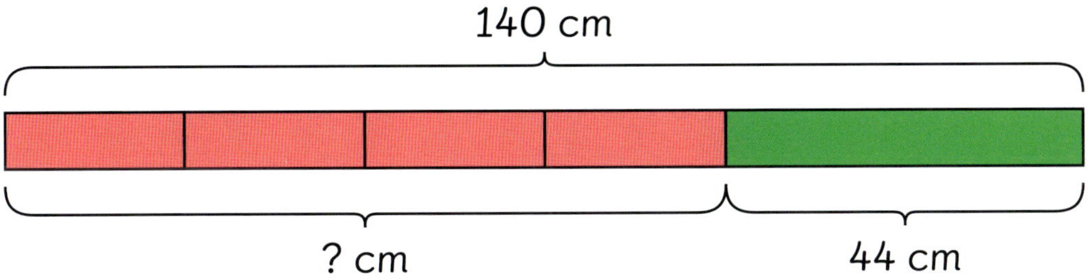

140 − 44 = 96

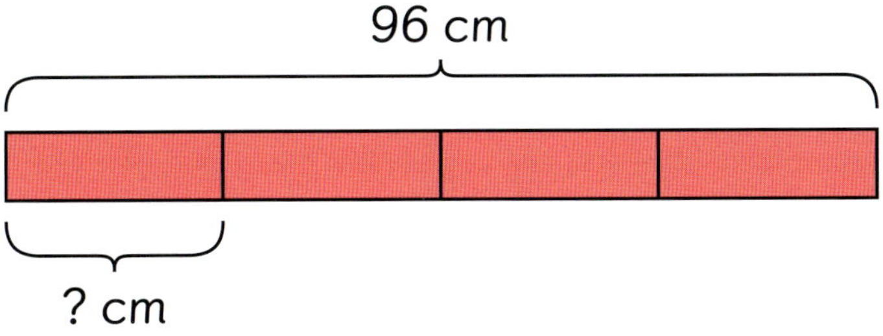

96 ÷ 4 = 24

Holly used 24 cm of wrapping paper to wrap each present.

Guided Practice

1 Charles has 490 cm of thread.
He makes 8 bracelets from it.
Each bracelet uses 52 cm of thread.
How much thread does Charles have left?

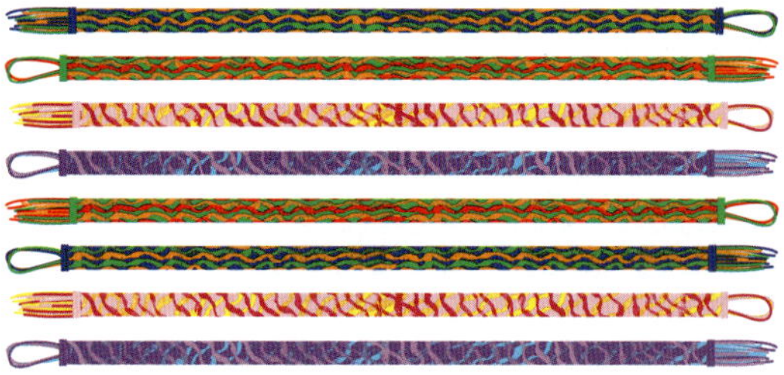

2 A machine slices a 5-cm thick block of wood into several 8-mm thick panels.

(a) How many 8-mm panels can we get?

(b) What is the thickness of the remaining block of wood?

3 Ravi's dad is 3 cm taller than Ravi's mum. Ravi's mum is twice as tall as Ravi. Ravi is twice as tall as his dog. Ravi's dad is 183 cm tall.
How tall is Ravi's dog?

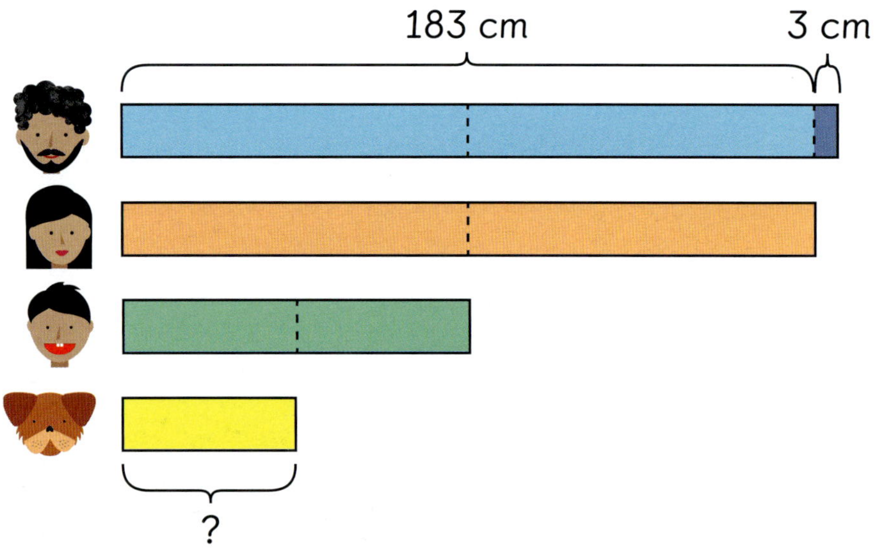

Complete Worksheet **10**: pages **161–162**

Mind Challenge

Sam's class grew a sunflower from a seed. They measured the height of the sunflower once a month.
In the first month, its height was 78 cm.
In the second month, its height had increased by 34 cm.
In the third month, the height had increased twice as much as it did during the second month.
What was the height of the sunflower in metres and centimetres in the third month?

My Maths Journal

Find an example where a measurement has been given as a single unit of measurement and another example where it has been given as mixed units of measurement.

For example:

The News

A grass snake found in the UK can grow up to 150 cm

The News

An albatross with a 3 m 70 cm wingspan has been seen off the coast of New Zealand

Write your own headline using a single unit of measurement and another headline using mixed units of measurement.

Review

I can...

- [ ] measure and write length in centimetres (cm), or in metres (m) and centimetres (cm).
- [ ] convert length from centimetres to metres and centimetres.
- [ ] convert length from metres and centimetres to centimetres.
- [ ] measure and write length in millimetres (mm), or in centimetres (cm) and millimetres (mm).
- [ ] convert length from millimetres to centimetres and millimetres.
- [ ] convert length from centimetres and millimetres to millimetres.
- [ ] compare the lengths of different objects.
- [ ] solve word problems involving length using a bar model.

Which is the heaviest item?

Chapter 6
Mass

Reading Weighing Scales

Lesson 1

Explore

Which item is the heaviest and which is the lightest?

Master

Use a scale to find the mass.

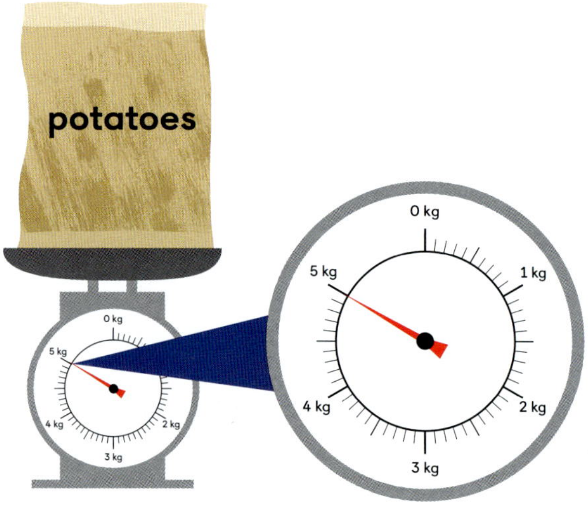

The mass of the bag of potatoes is about 5 kg.

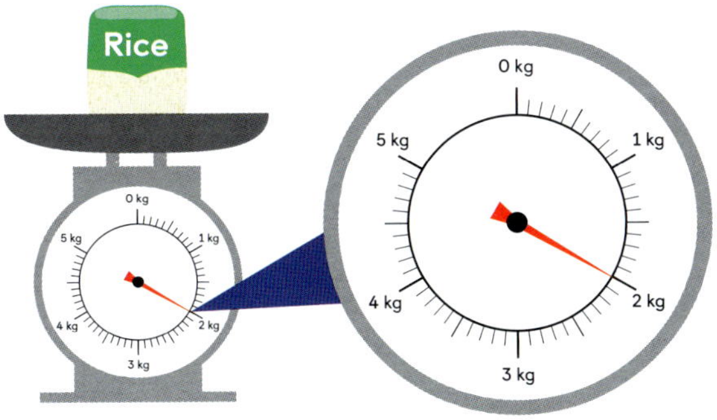

The mass of the bag of rice is about 2 kg.

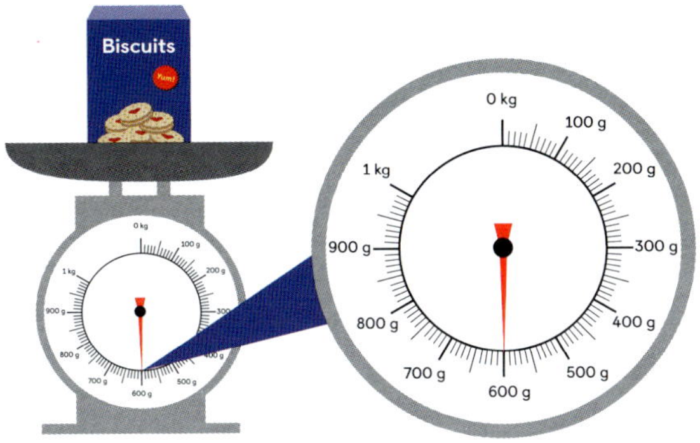

The mass of the box of biscuits is about 600 g.

The bag of rice has a greater mass than the box of biscuits.

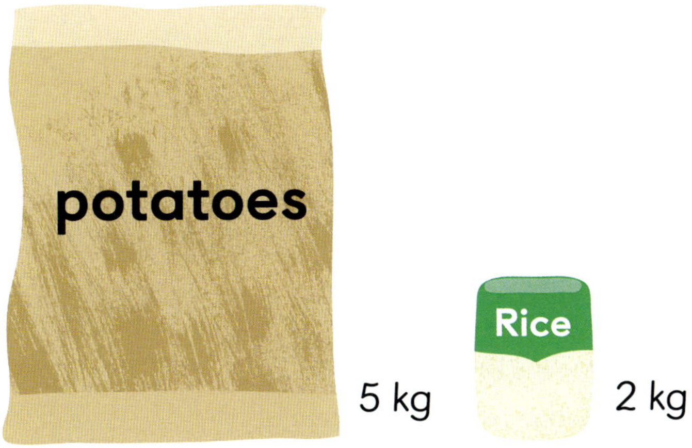

The bag of potatoes has a greater mass than the bag of rice.

The bag of potatoes has the greatest mass. It is the heaviest.
The box of biscuits has the smallest mass. It is the lightest.

Guided Practice

These items are weighed on different scales.
What is the mass of each item?

1

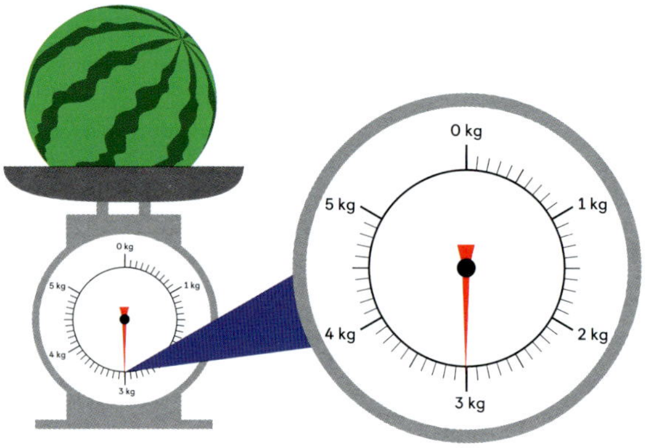

The mass of the watermelon is about ☐ kg.

2

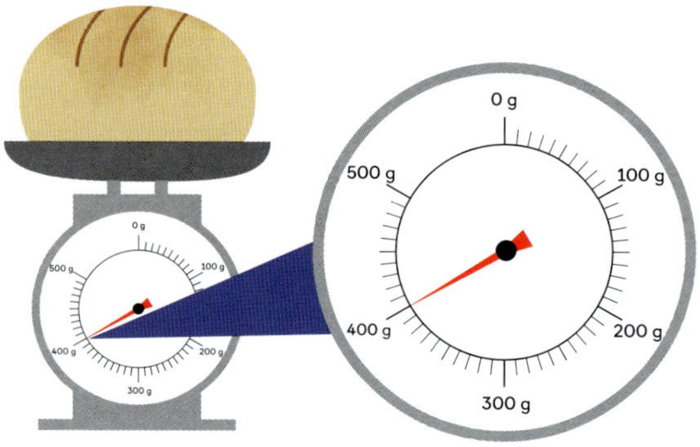

The mass of the loaf of bread is about ☐ g.

3

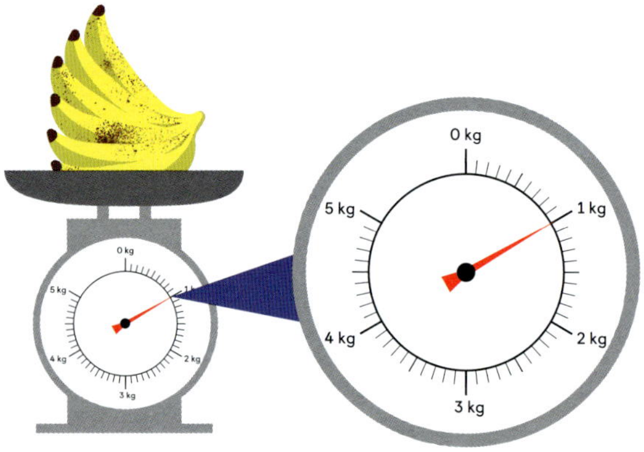

The mass of the bunch of bananas is about ☐ kg.

4

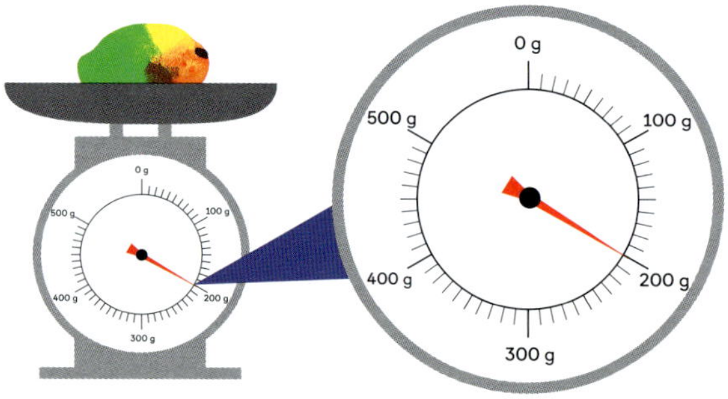

The mass of the mango is about ☐ g.

Complete Worksheet **1**: pages **169–170**

Reading Weighing Scales

Lesson 2

Explore

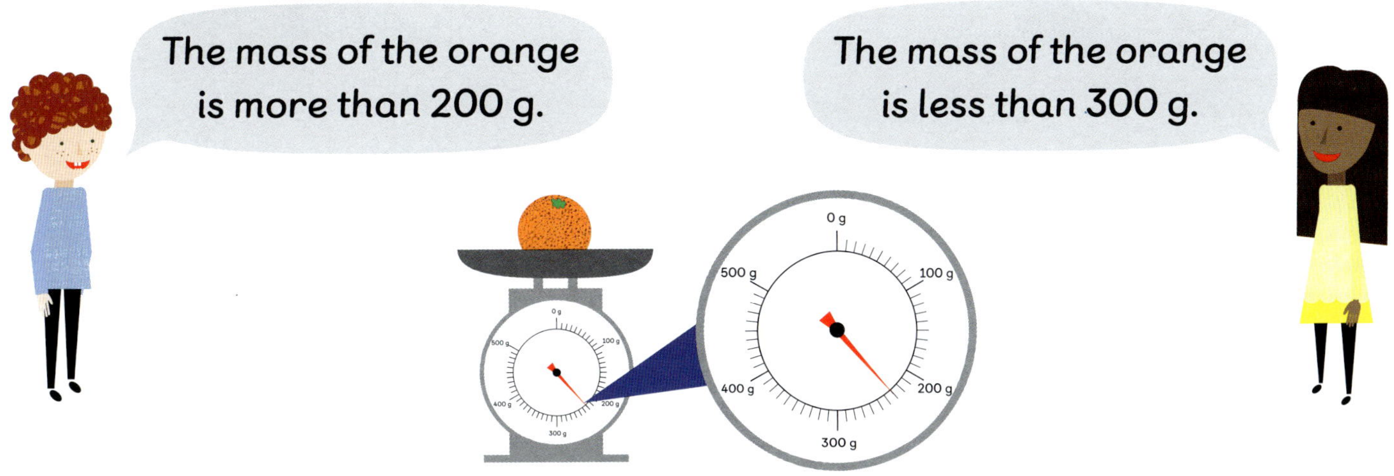

What is the mass of the orange?

Master

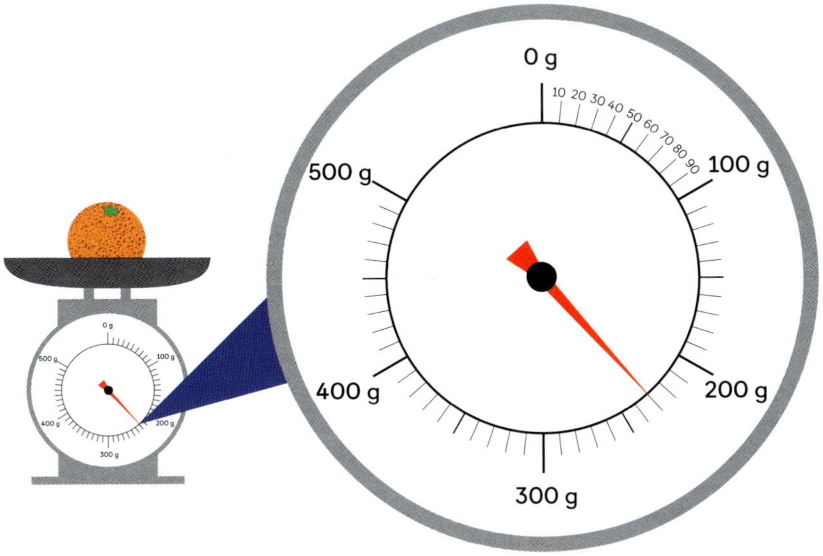

The mass of the orange is about 230 g.

Guided Practice

Find the mass of the items.

1

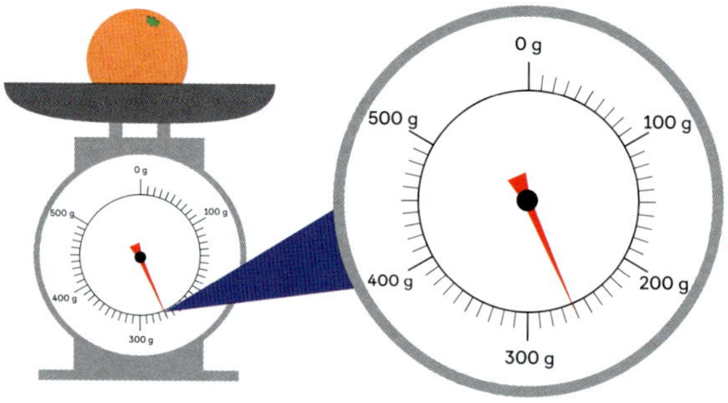

The mass of the grapefruit is about ______ g.

2

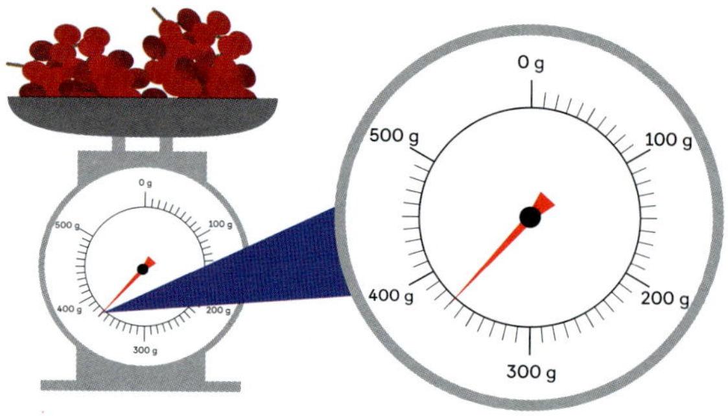

The mass of the grapes is about ______ g.

3

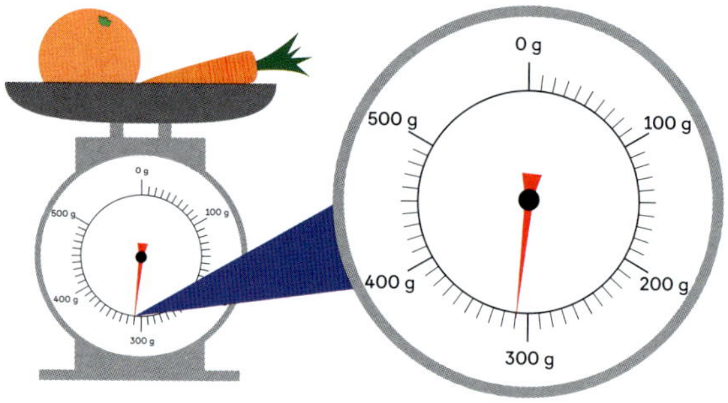

The mass of the grapefruit and the carrot is about ______ g.

4

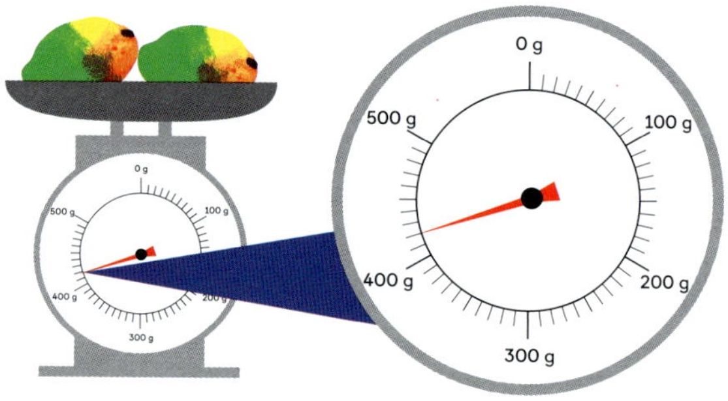

The mass of the 2 mangoes is about ______ g.

5 Fill in the blanks with **greater** or **smaller**.

The mass of 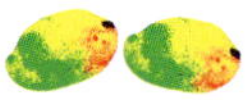 is ______ than the mass of the .

The mass of  is ______ than the mass of the 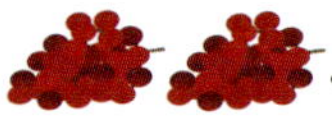.

Complete Worksheet **2**: pages **171–172**

Reading Weighing Scales

Lesson 3

Explore

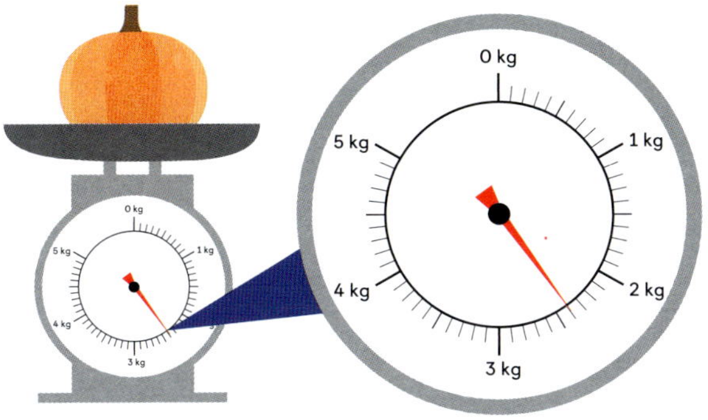

The pumpkin has a mass greater than than 2 kg and smaller than 3 kg.

What is the mass of the pumpkin?

Master

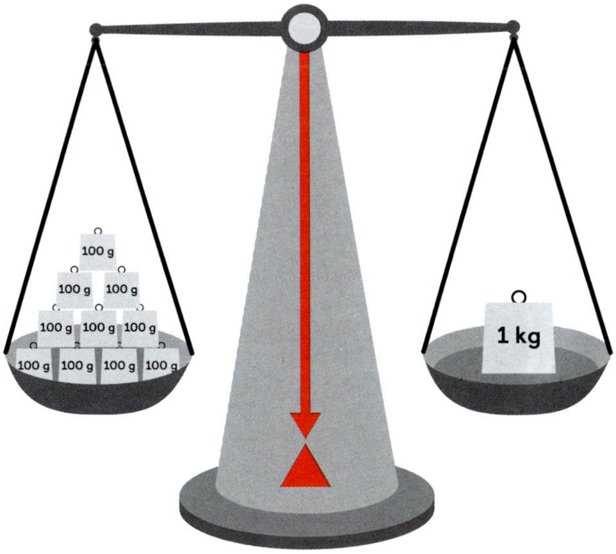

A one-kilogram mass is equal to ten 100-gram masses.

1 kg = 10 × 100 g

1 kg = 1000 g

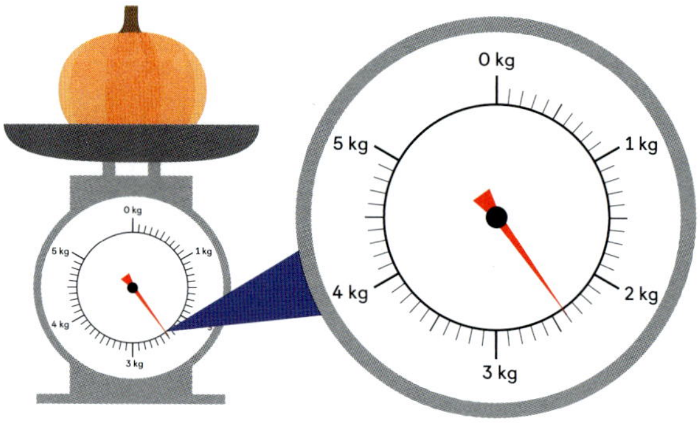

The mass of the pumpkin is about 2 kg 400 g.

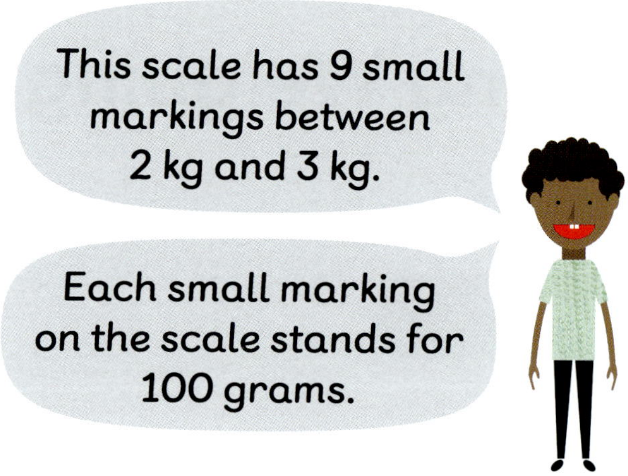

Guided Practice

Read the scales to find the mass of each item.

1

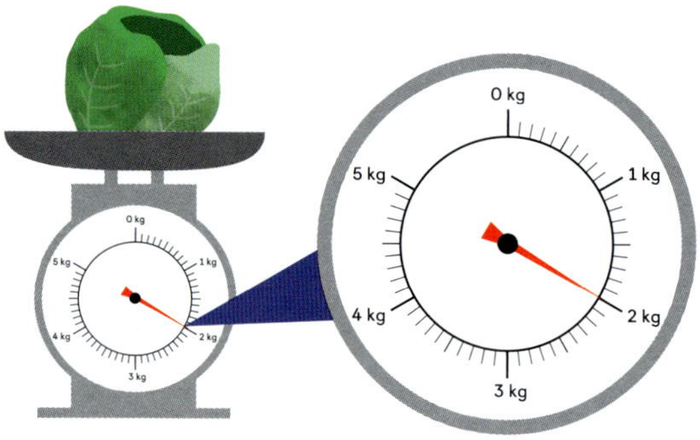

The mass of the cabbage is about ______ kg.

2

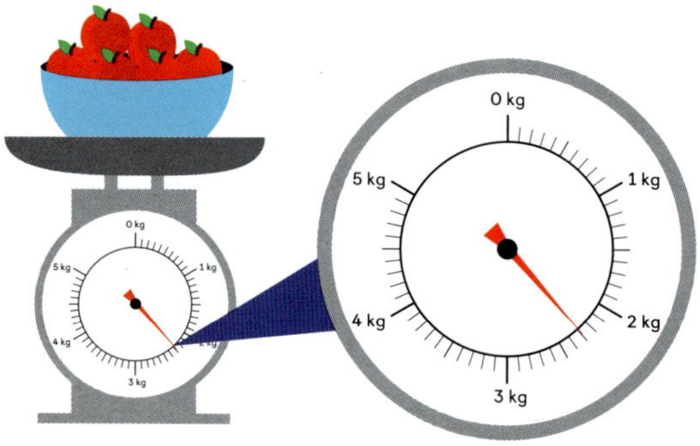

The mass of the bowl of apples is about ______ kg 300 g.

3

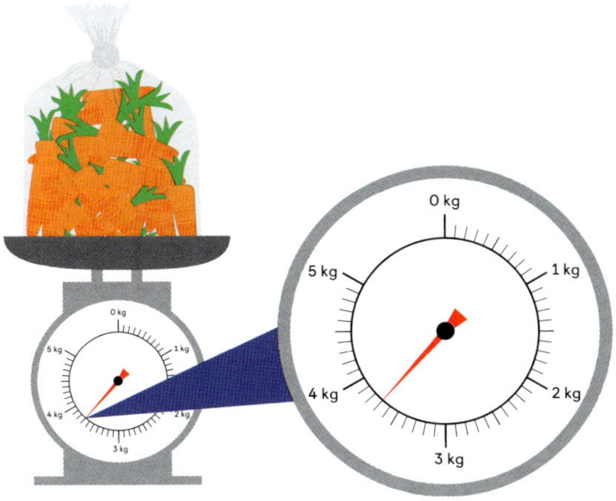

The mass of the bag of carrots is about 3 kg ____ g.

4

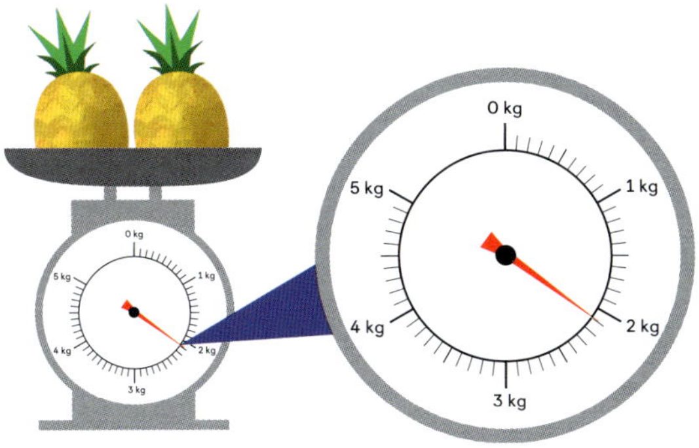

The mass of the 2 pineapples is about ____ kg ____ g.

5

Fill in the blanks with **greater** or **smaller**.

The mass of  is ____ than the mass of the .

The mass of  is ____ than the mass of the .

Complete Worksheet **3**: pages **173–174**

Reading Weighing Scales

Lesson 4

Explore

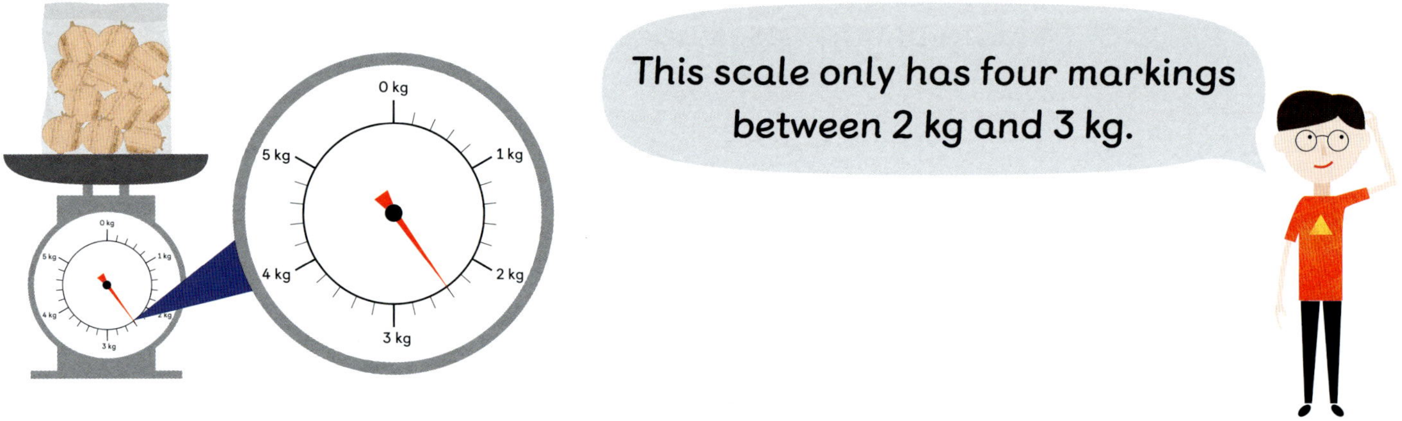

What is the mass of the bag of onions?

Master

1

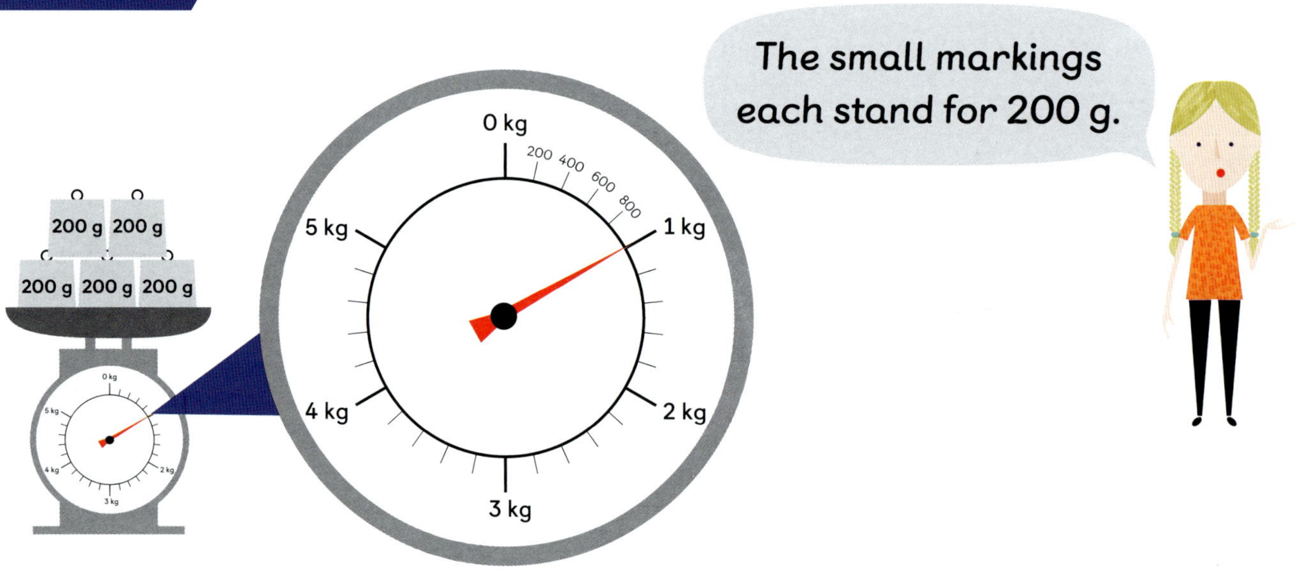

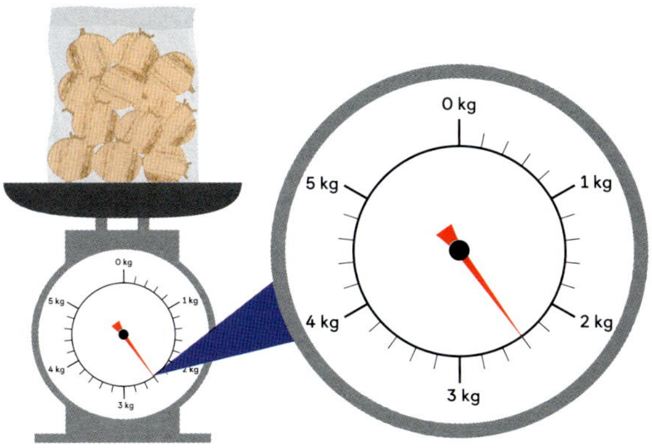

The mass of the bag of onions is about 2 kg 400 g.

2 What is the mass of each item?

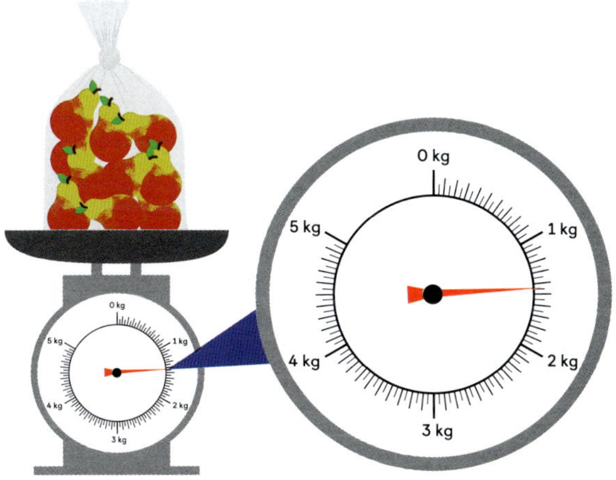

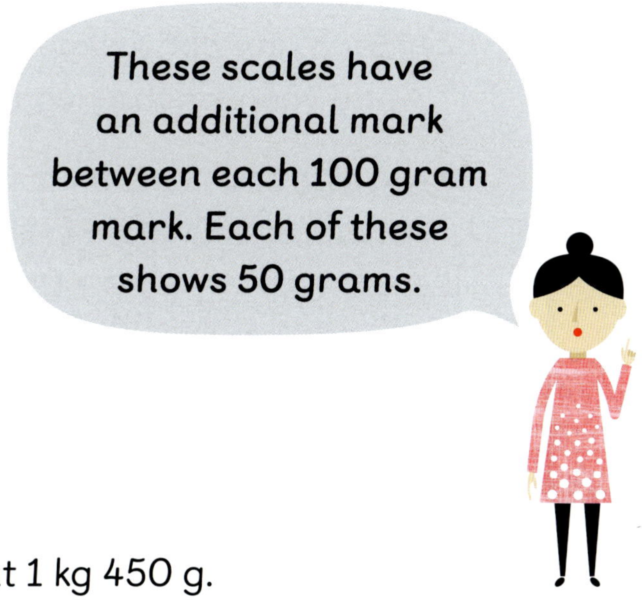

The mass of the bag of pears is about 1 kg 450 g.

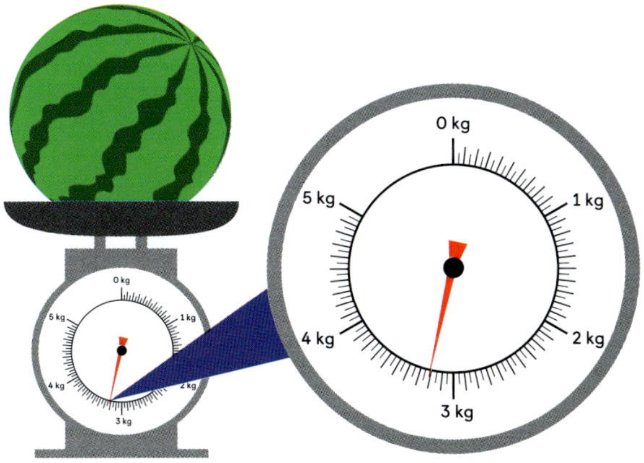

The mass of the watermelon is about 3 kg 200 g.

Activity Time

Partner work

You will need:

(1) Each partner finds an object in your classroom:

(a) that you estimate has a mass of less than 1 kg.

(b) that you estimate has a mass of more than 1 kg.

(c) that you estimate has a mass of about 2 kg.

(2) Each partner estimates the mass of all the objects and records their estimates in a table.

(3) Weigh the items on a scale and record the actual mass of each item in the table.

Object	Partner 1's estimate	Partner 2's estimate	Actual mass
water bottle	less than 1 kg	more than 1 kg	750 g

(4) Compare your estimates with the actual mass of each object.

Guided Practice

How heavy is each item?

1

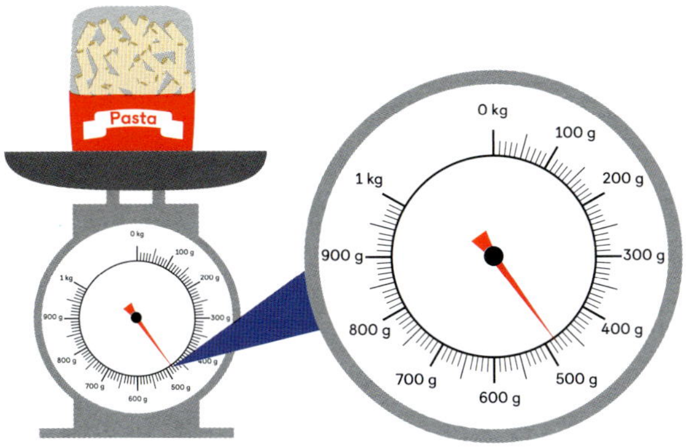

The bag of pasta has a mass of about ▭ g.

2

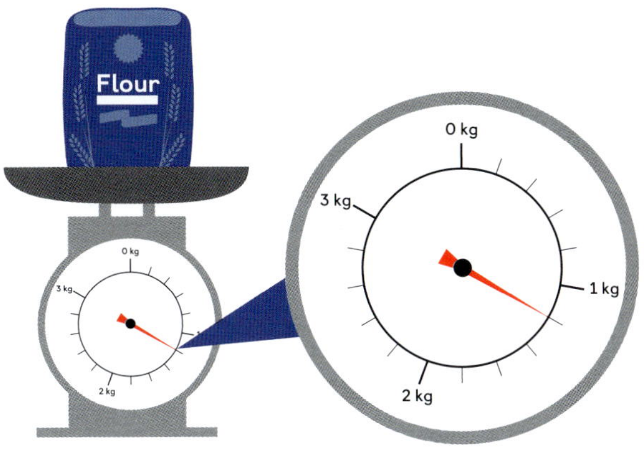

The bag of flour has a mass of about ▭ kg ▭ g.

3

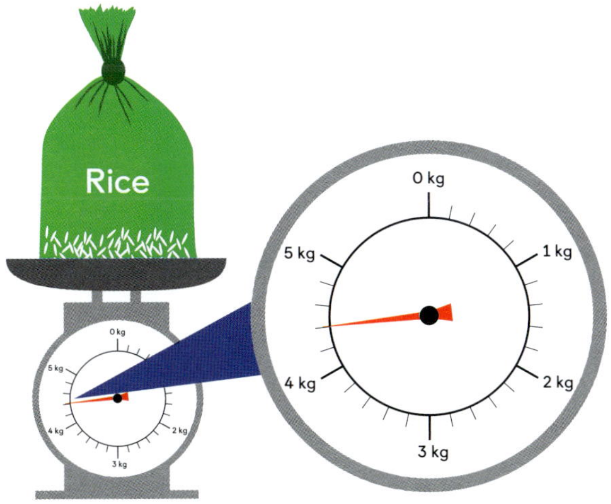

The bag of rice has a mass of about ▭ kg ▭ g.

4

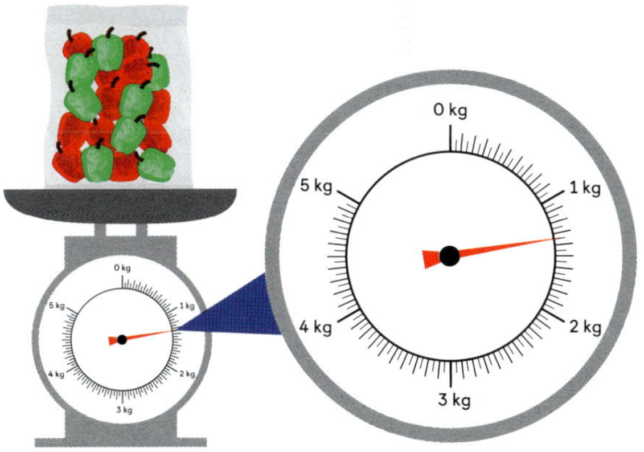

The bag of peppers has a mass of about ▭ kg ▭ g.

Complete Worksheet 4: pages 175–176

Solving Word Problems

Lesson 5

Explore

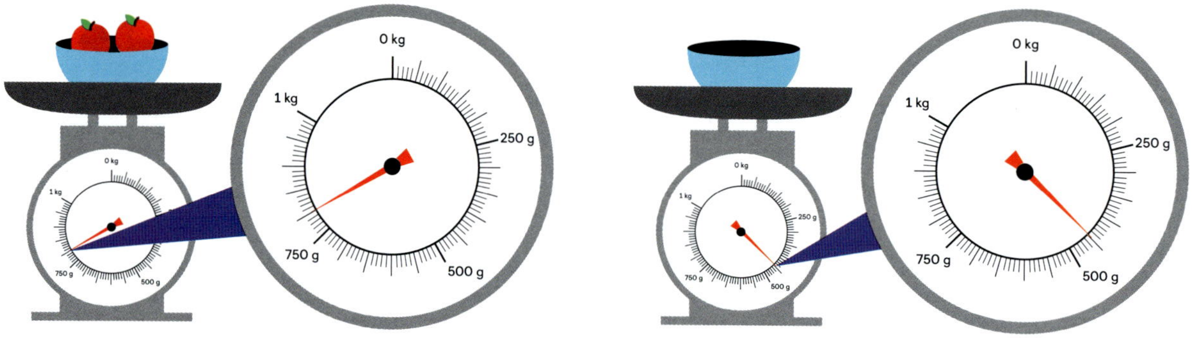

What is the mass of the apples?

How many grams do the small markings show on this scale?

Master

The mass of the bowl of apples is about 800 g.
The bowl is about 450 g.

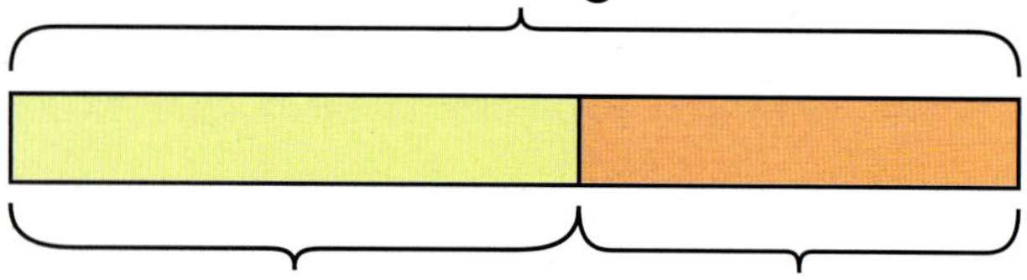

	8	0	0
−	4	5	0
	☐	☐	☐

800 − 450 = 350

The mass of the apples is about 350 g.

Guided Practice

1 Holly has 735 g of flour.
She uses 545 g to make pizza dough.
Find the mass of the flour she has left.

2 A box of P.E. equipment has a mass of 11 kg.
The empty box has a mass of 3 kg.
Find the mass of the P.E. equipment.

3 Find the total mass of the pineapples and bananas.

4 Find the total mass of the watermelon and the bag of oranges.

Complete Worksheet **5**: pages **177–178**

Solving Word Problems

Lesson 6

Explore

Emma is baking bread with her aunt. They use 15 g of yeast for the first loaf. They want to bake 3 more identical loaves. How much yeast do they need in total for all the loaves?

Master

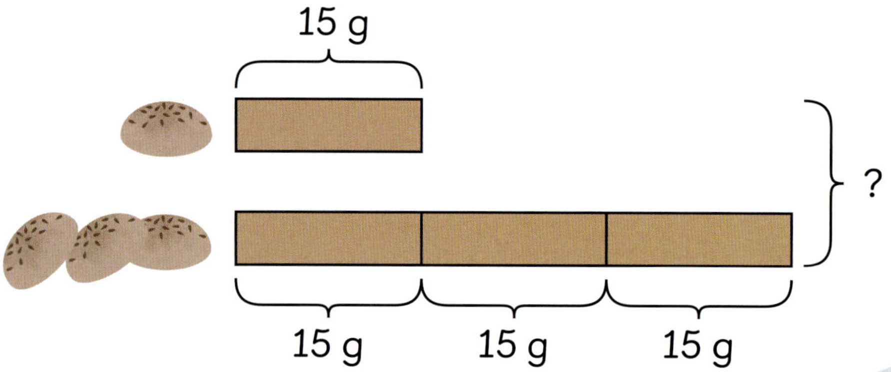

	1	5
×		4
	2	0
+	4	0
	☐	☐

15 × 4 = 60

Emma and her aunt need 60 g of yeast to bake 4 loaves of bread.

Guided Practice

1 Elliott used 67 g of butter to make some cookies.
Emma used twice as much butter to make her cookies.
How much butter did Emma use?

2 An apricot has a mass of 35 g.
An orange has 4 times the mass of an apricot.
What is the total mass of the 2 fruits?

3 A grapefruit has a mass 5 times the mass of a kiwifruit. The mass of the kiwifruit is 50 g.
What is the total mass of 3 grapefruits and 2 kiwifruits?

Complete Worksheet **6**: pages **179–180**

Lesson 7

Solving Word Problems

Explore

A worker puts 5 identical new televisions on a pallet. The total mass of the pallet and the televisions is 82 kg.
If the pallet has a mass of 17 kg, what is the mass of each television?

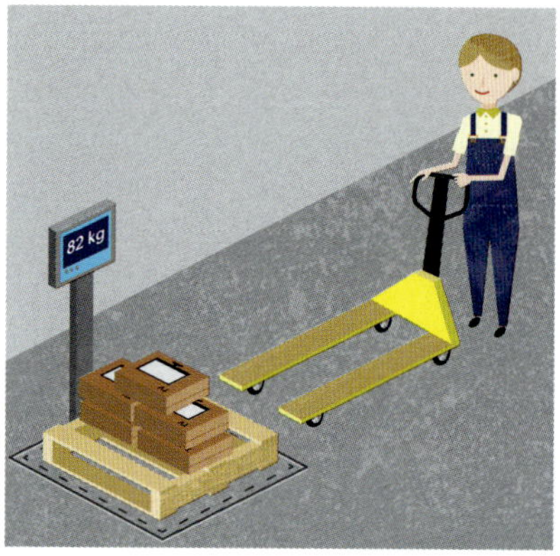

Master

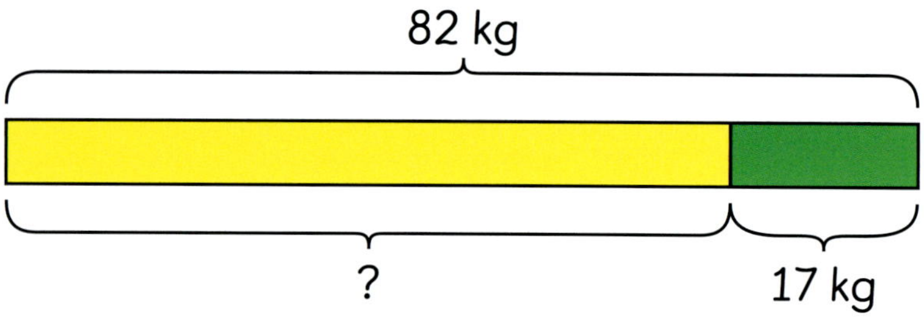

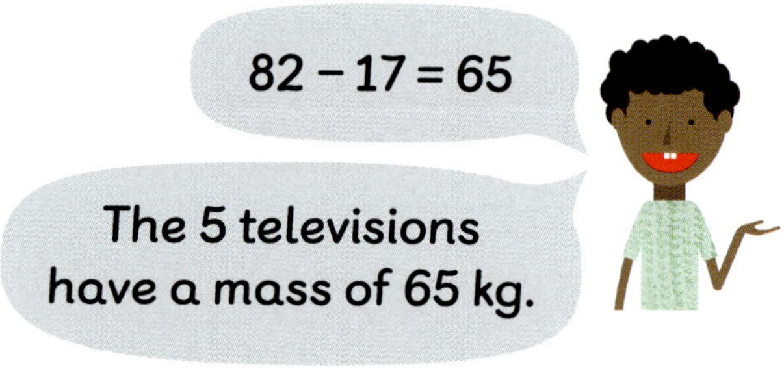

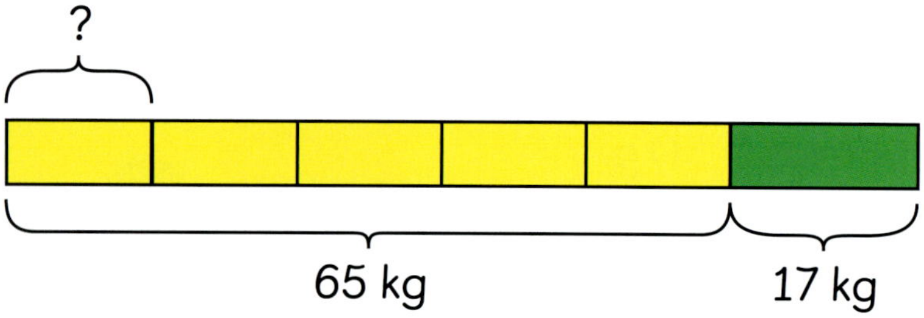

$$\begin{array}{r|rr} & 1 & 3 \\ \hline 5 & 6 & 5 \\ - & 5 & 0 \\ \hline & 1 & 5 \\ - & 1 & 5 \\ \hline & & 0 \end{array}$$

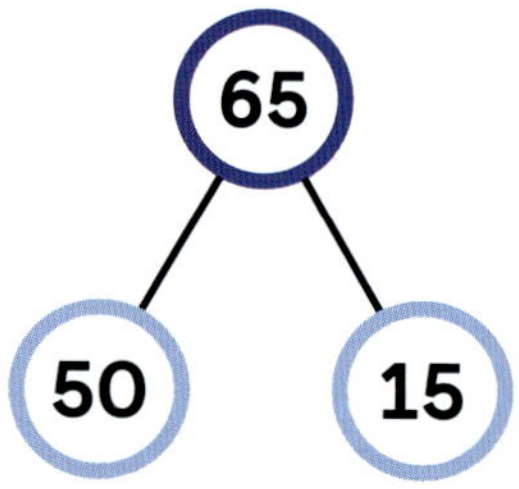

Each television has a mass of 13 kg.

Guided Practice

1 Hannah fills 5 bowls with an equal amount of granola.
After filling the bowls, there is 95 g of granola left.
What is the mass of granola in each of the bowls?

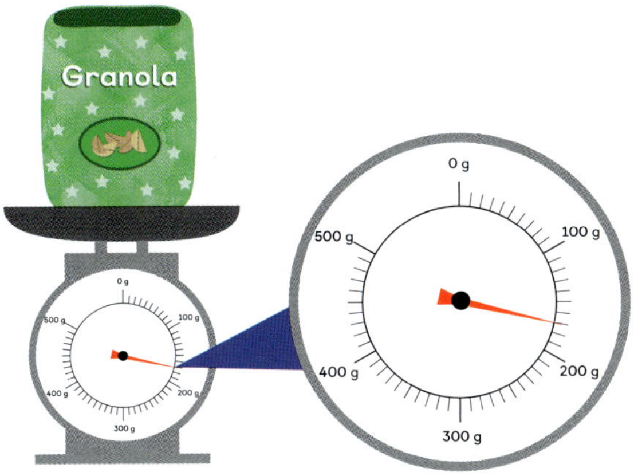

2 A tray with 4 identical calculators in it has a mass of 600 g.
After 3 calculators are taken out, the tray and the remaining calculator have a mass of 360 g.
What is the mass of each calculator?

3 The mass of a box with 18 identical packets of crisps is 550 g.
After 8 packets are taken away, the mass of the box and the remaining packets of crisps is 350 g.
What is the mass of each packet of crisps?

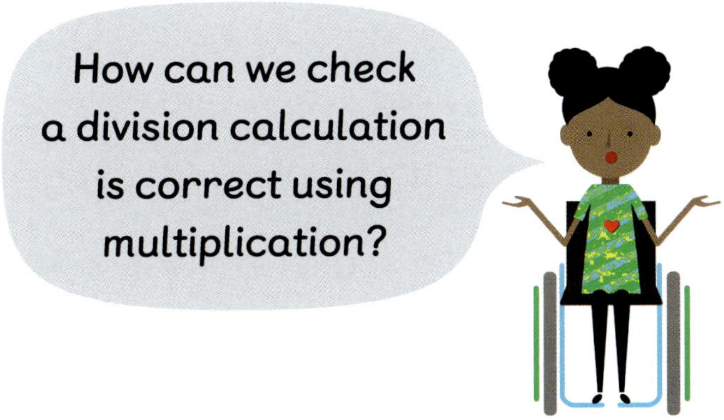

Complete Worksheet **7**: pages **181–182**

Mind Challenge

A package with a mass of more than 12 kg is labelled as 'heavy'.
5 pots need to be sent in a package.
Each pot has a mass of 2 kg 125 g and the empty package has a mass of 700 g.
Will the package be labelled as 'heavy'?

My Maths Journal

Where is mass measured on a daily basis?
Give three examples and explain why.

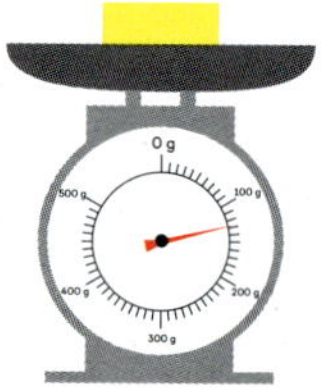

I can...

Review

- ☐ read a scale for mass in grams (g).
- ☐ read a scale for mass in kilograms (kg).
- ☐ read a scale for mass in kilograms and grams.
- ☐ solve word problems involving mass.

What is the volume of liquid in each container?

Chapter 7
Volume

Measuring Volume in Millilitres

Lesson 1

Explore

How much orange juice is in each container?
How can we find out?

Master

1 We can measure the amount of orange juice by pouring it into a measuring beaker.

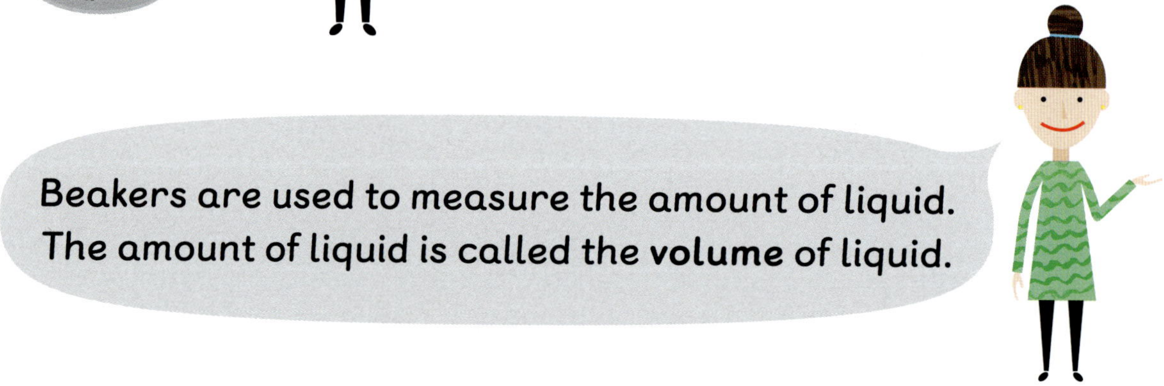

2

This beaker has markings for each 50 millilitres or 50 ml.

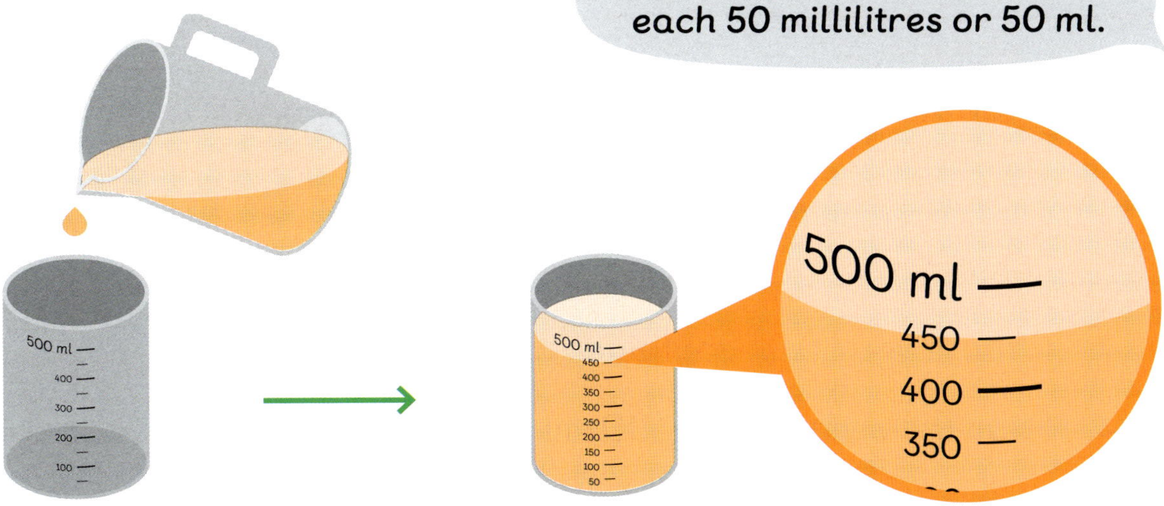

The volume of orange juice in the jug is about 450 ml.

ml stands for millilitres.

3

The glass has less orange juice in it. We can use the smaller beaker to measure its volume.

The smaller beaker has markings for each 10 ml.

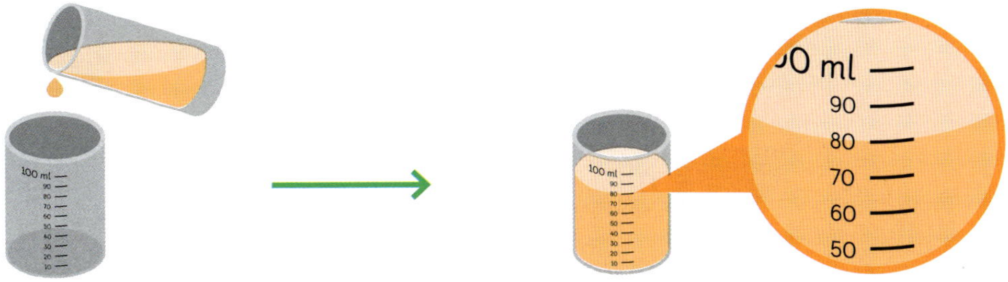

The volume of juice in the glass is about 80 ml.

Activity Time

Partner work

You will need:

1. Find three or four small containers.
2. Choose your first container and fill it with water.
3. Pour the water into a 100-ml beaker.
4. Measure the volume of water in the beaker and write it down.
5. Repeat with each of the containers in turn.
6. Arrange the containers in order, based on the volume of water they can hold.

Guided Practice

Find the volume of water in each beaker.

1

☐ ml

2

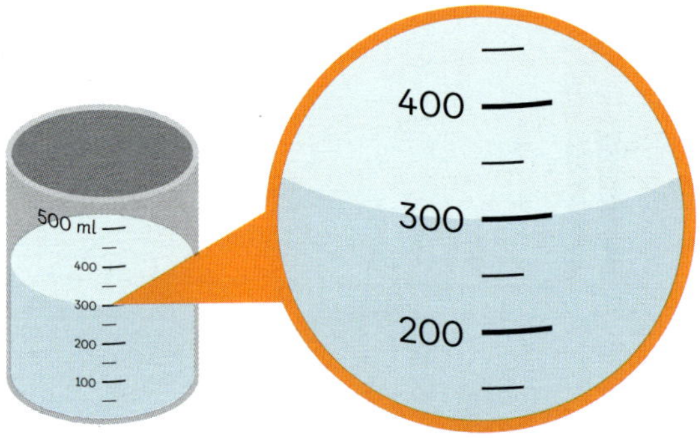

☐ ml

3

ml

4

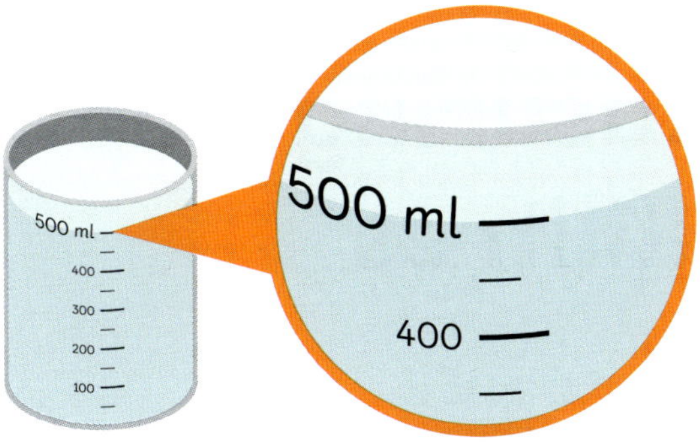

ml

5

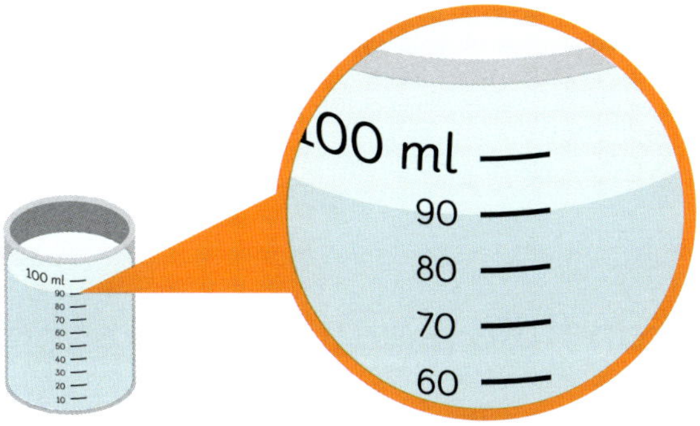

ml

6

ml

7

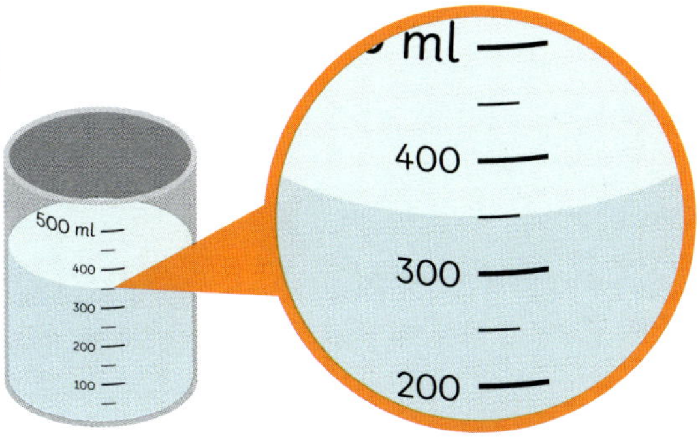

ml

8

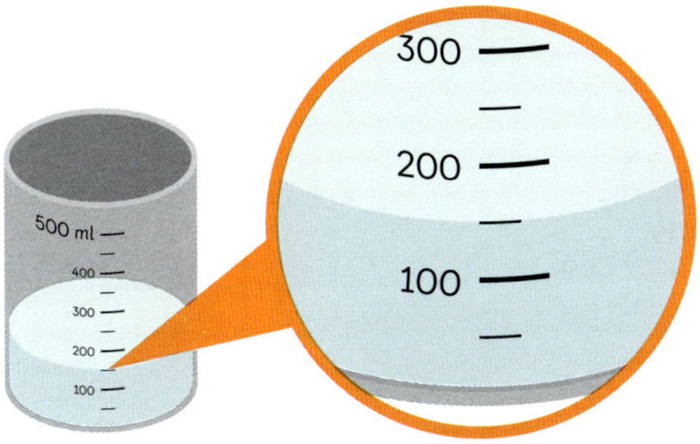

ml

Complete Worksheet **1**: pages **187–188**

Measuring Capacity in Millilitres

Lesson 2

Explore

Hannah wants to store her leftover soup in the larger one of these two containers. Which of the two containers can hold the most soup?

Master

The volume of liquid a container can hold is called its **capacity**.

We can use a  to measure the capacity of each container.

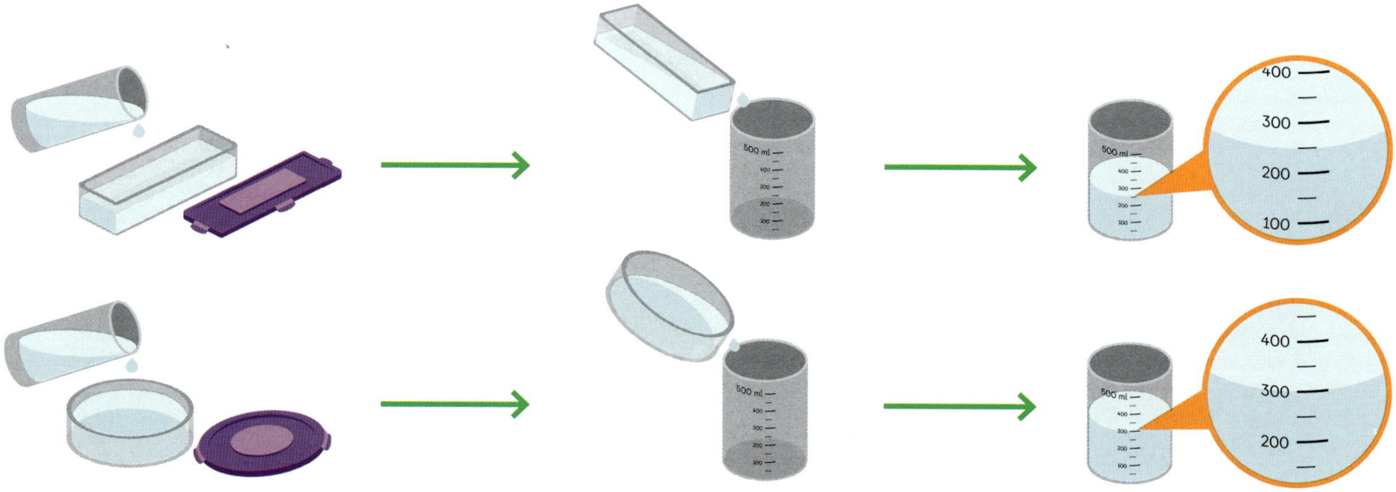

 has a capacity of 250 ml.

 has a capacity of 300 ml.

 has a larger capacity. It can hold the most soup.

Activity Time

Play a game (3 to 4 pupils)

You will need:

1. Each player chooses a container they think can hold about 500 millilitres of water.
2. Check the capacity of each container using the 500-millilitre beaker.
3. The winner is the player with the container that can hold the closest to 500 millilitres of water.
4. Play again using different containers.

Guided Practice

What is the capacity of each container?

1

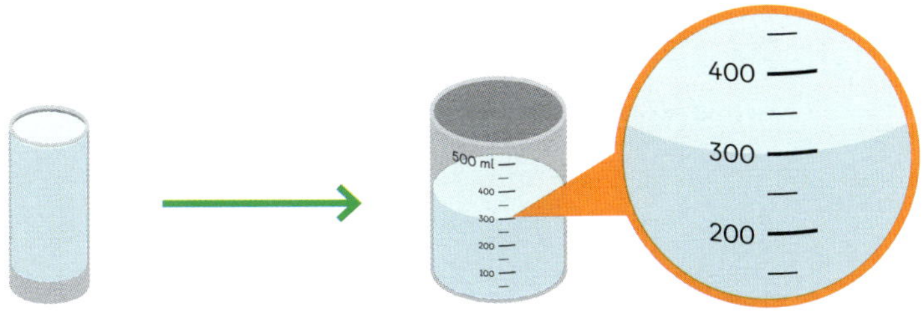

The capacity of the glass is ______ ml.

2

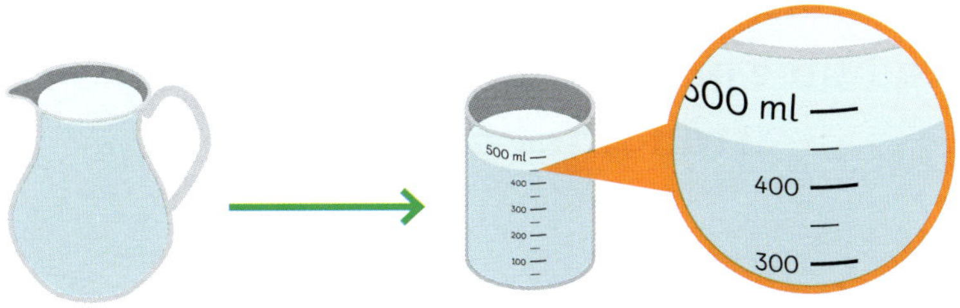

The capacity of the jug is ______ ml.

Complete Worksheet **2**: pages **189–190**

Measuring Volume in Millilitres and Litres

Lesson 3

Explore

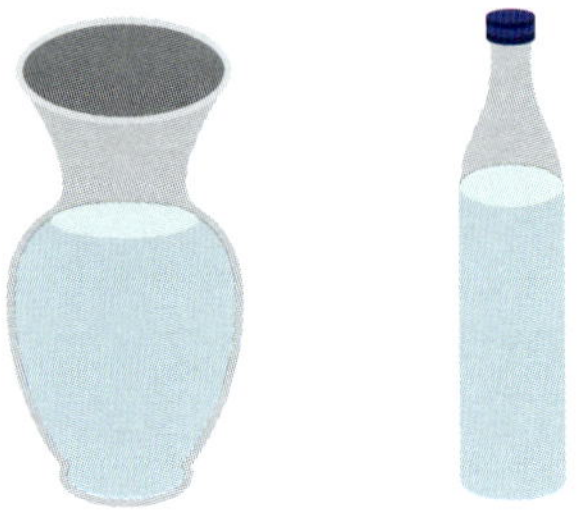

The vase and the bottle are not full.

I think there is 1 litre of water inside the vase and 750 millilitres of water in the bottle.

Is Sam correct? How can we find out?

Master

1

We can use these beakers to measure the volume of water inside the vase and the bottle.

Each of these beakers can measure up to 1 litre or 1 l. Each of the markings stands for 50 ml.

l stands for litres.

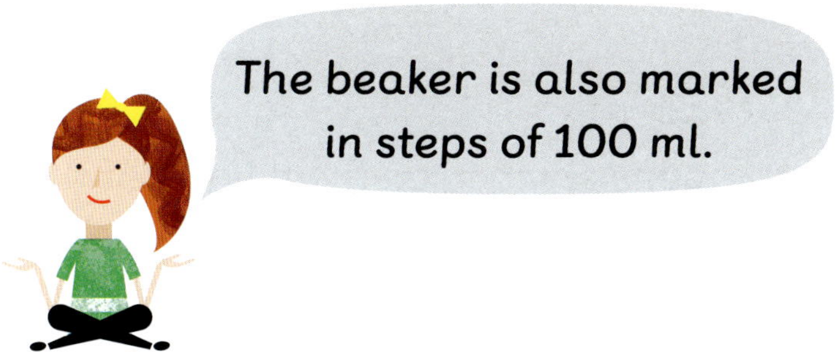

2

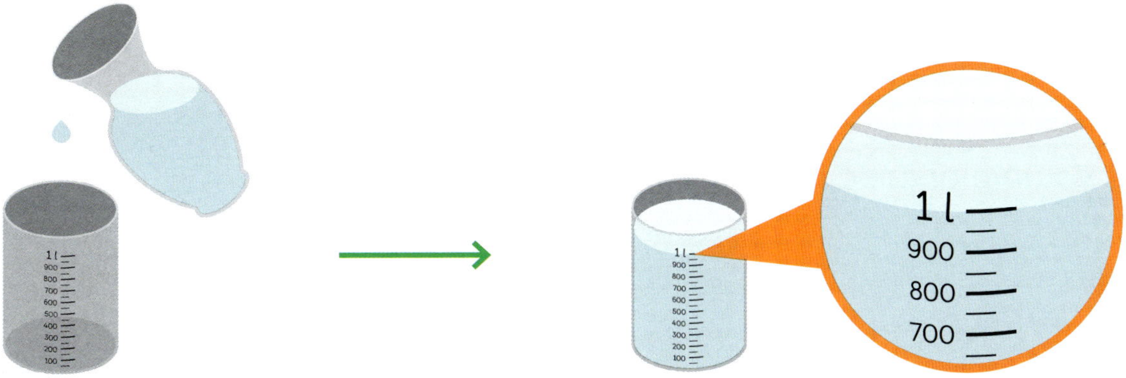

The volume of the water in the vase is about 1 l.

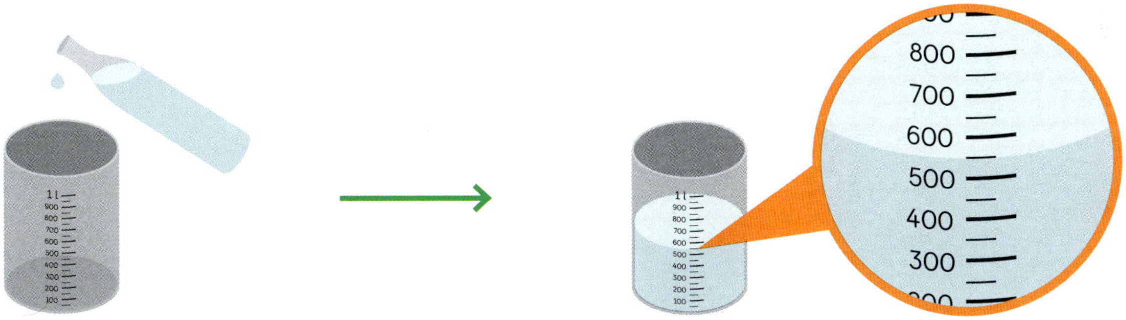

The volume of water in the bottle is about 550 ml.

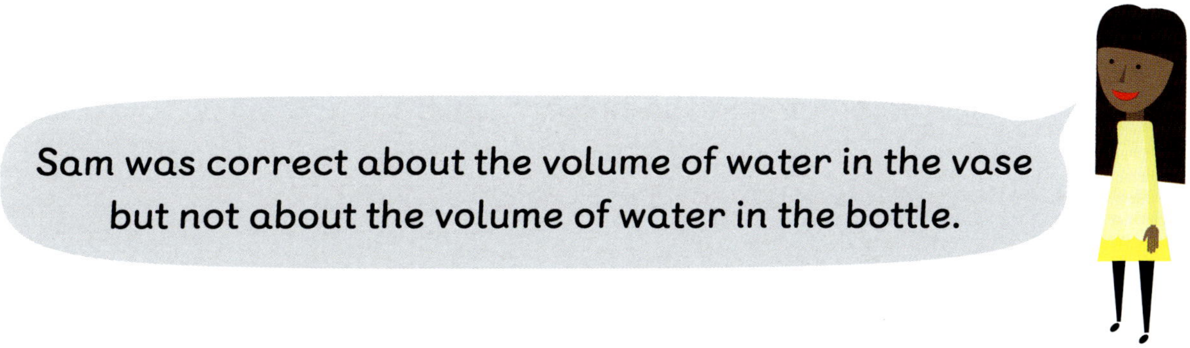

Guided Practice

Sam pours some juice from each of the bottles into 1-l beakers.
What is the volume of juice Sam pours into each beaker?

1

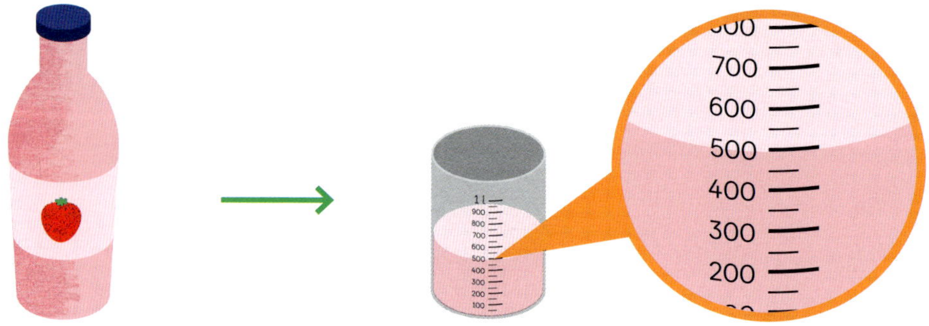

The volume of juice in the beaker is about ☐ ml.

2

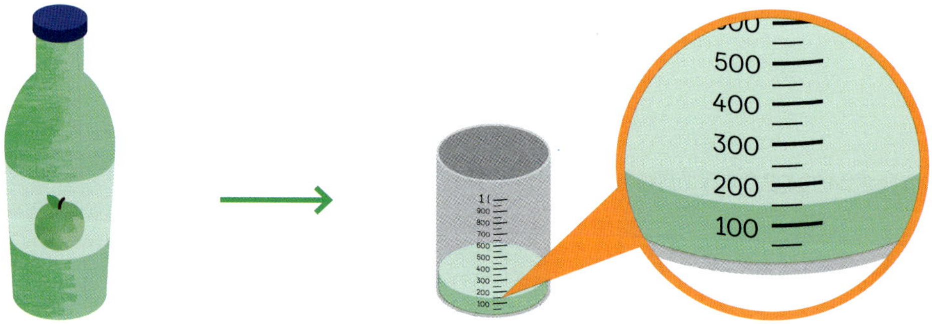

The volume of juice in the beaker is about ☐ ml.

3

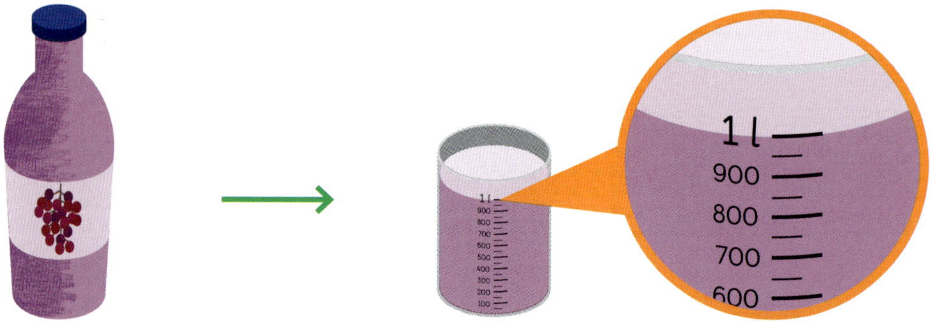

The volume of juice in the beaker is about ☐ l.

Complete Worksheet **3**: pages **191–192**

Measuring Capacity in Millilitres and Litres

Lesson 4

Explore

The label has come off Ruby's bottle.
How can she use a 100-ml beaker to find the capacity of her bottle?

Master

1 Ruby completes the following steps.

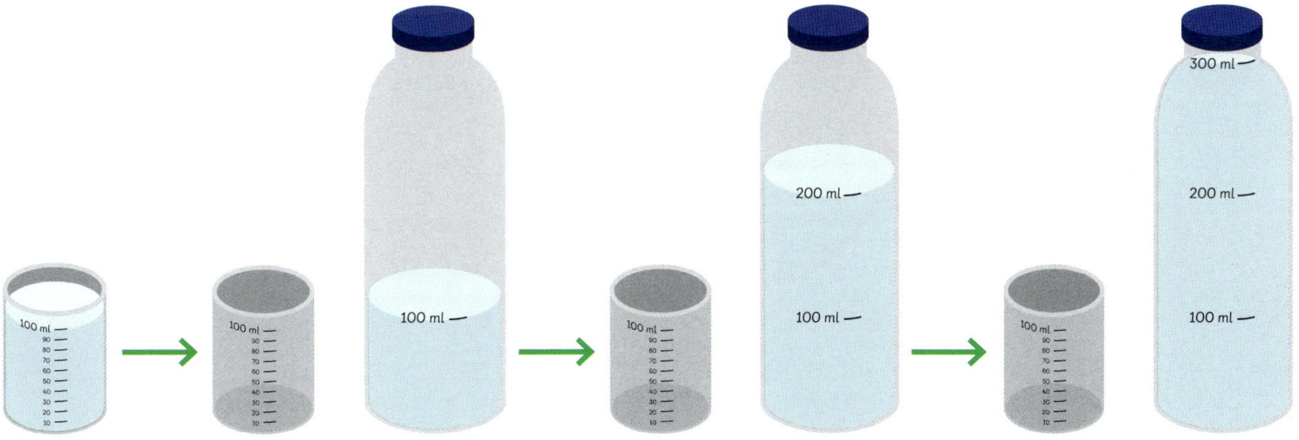

Step 1 Measure 100 ml of water using the 100-ml beaker.

Step 2 Pour the water into the bottle.

Step 3 Mark the water level on the bottle.

Step 4 Repeat steps 1 to 3 until the bottle is full.

The capacity of Ruby's bottle is 300 ml.

2 We can also measure the capacity of a container by pouring all the water from the container into a measuring beaker.

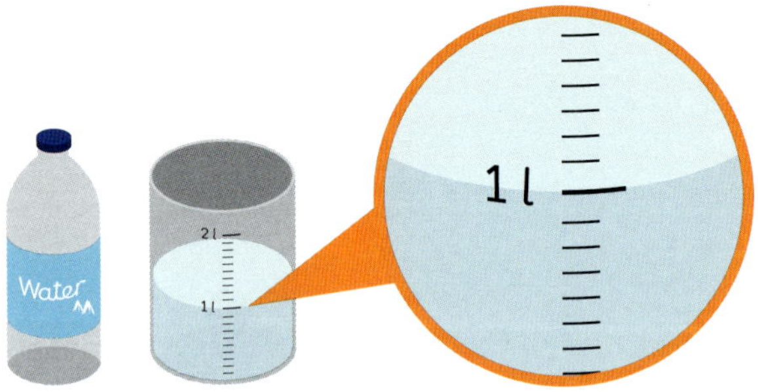

The capacity of the bottle is ____ l.

Guided Practice

1 Full containers are poured into different beakers.
What is the capacity of each container?

(a)

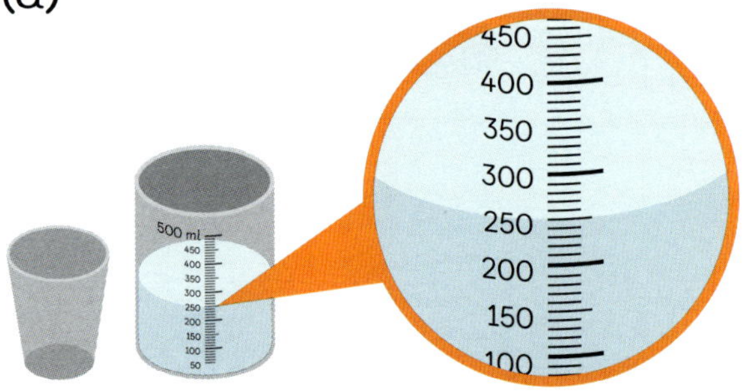

____ ml

(b)

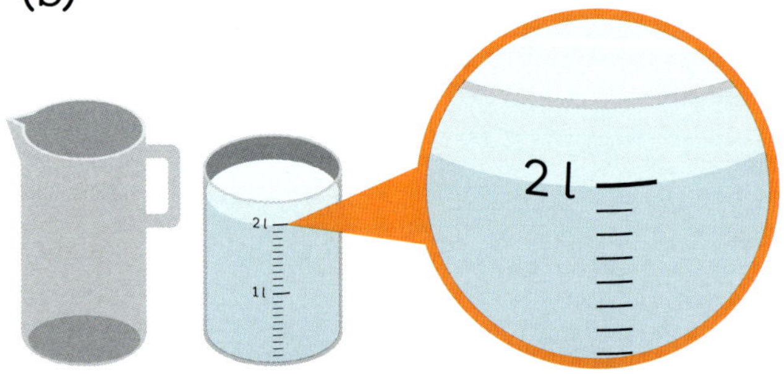

____ l

2 It takes three 250-ml glasses of water to fill a bottle.
What is the capacity of the bottle?

Complete Worksheet **4**: pages **193–194**

Writing Volume in Litres and Millilitres

Lesson 5

Explore

Jacob needs to put 1 l of water in the bucket.
Each glass has a capacity of 250 ml.
Each water bottle has a capacity of 500 ml.
Using the glasses and the bottles,
how can he put 1 l of water in the bucket?

Master

1

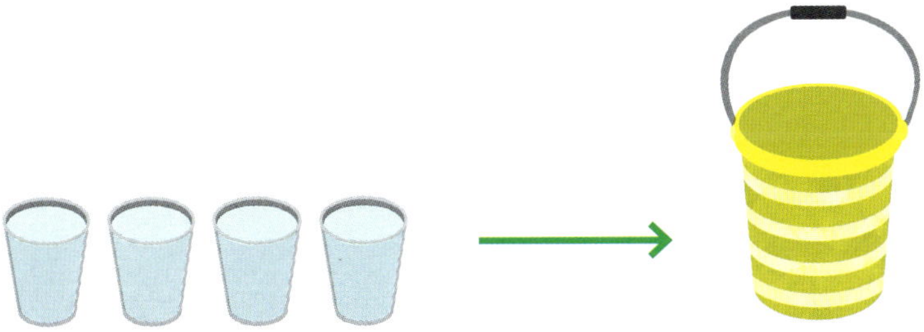

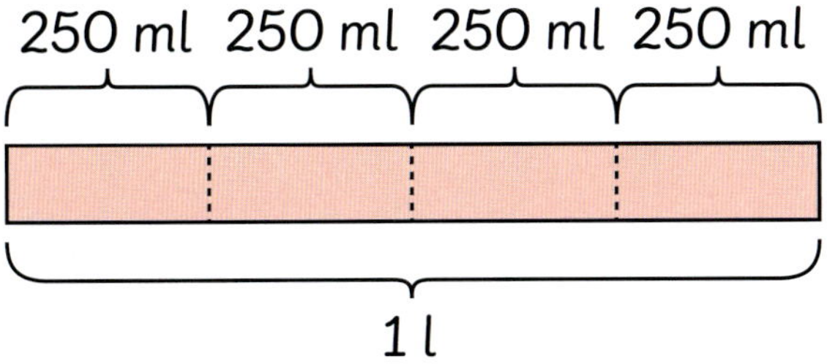

250 + 250 + 250 + 250 = 1000
4 glasses of water = 1000 ml

1000 ml = 1 l
The total volume of water in the 4 glasses is 1 l.

2

We can also pour water from 2 filled bottles into the bucket.
500 + 500 = 1000

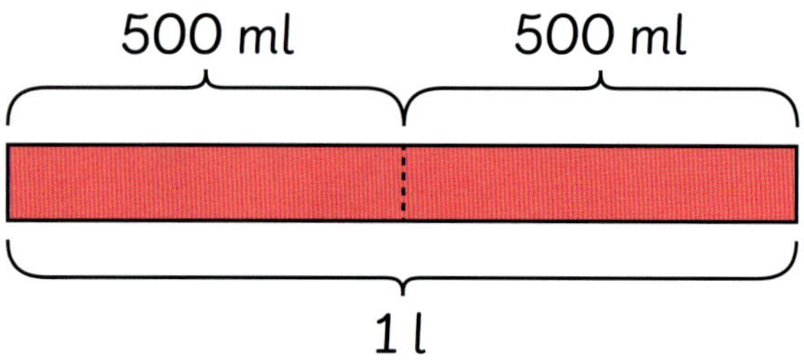

The total volume of water in the two water bottles is 1 l.

Are there more ways to pour 1 l of water into the bucket?

Guided Practice

1 Can you help  find the total volume of water in the two beakers?

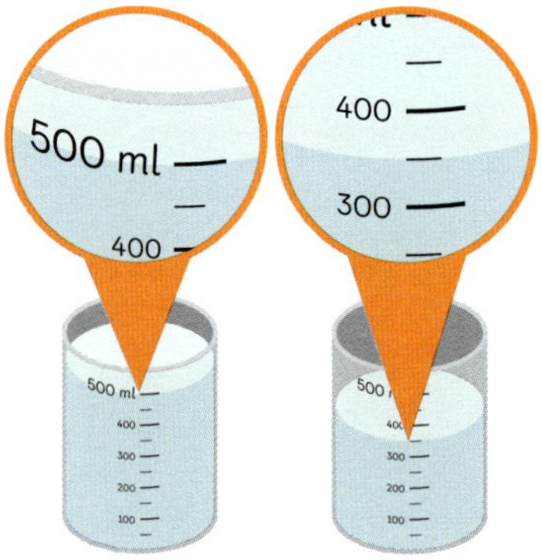

[ ] ml

2 How much more water does  need to put in the beaker to make 700 ml?

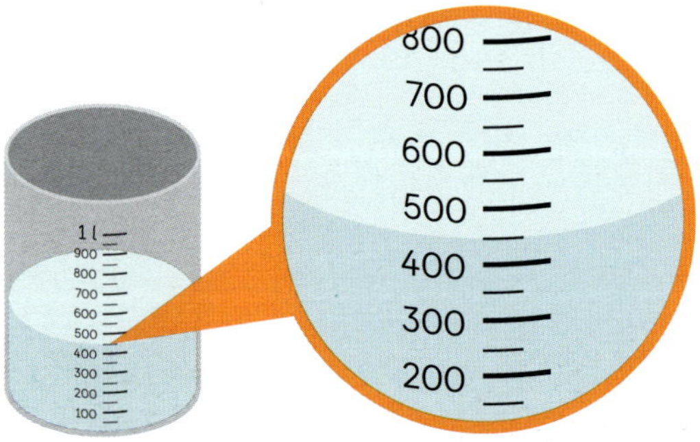

[ ] ml

3 If  uses 250 ml of water, how much water will remain in the beaker?

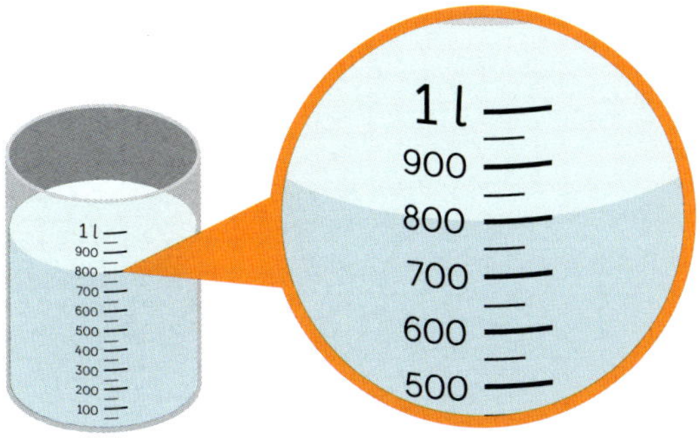

ml

4 Can you help Ravi find how much more water is needed to make 1 l?

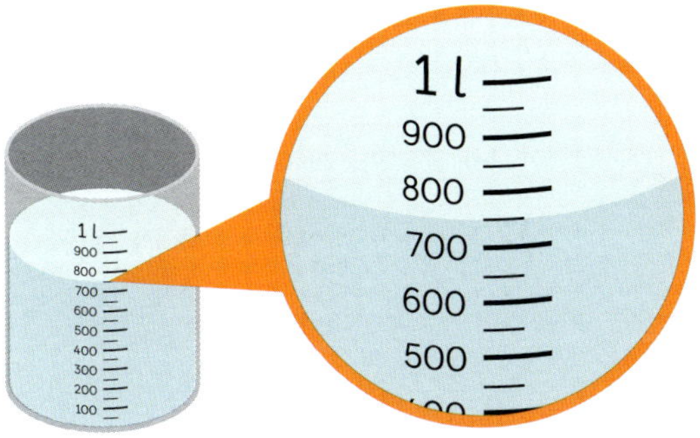

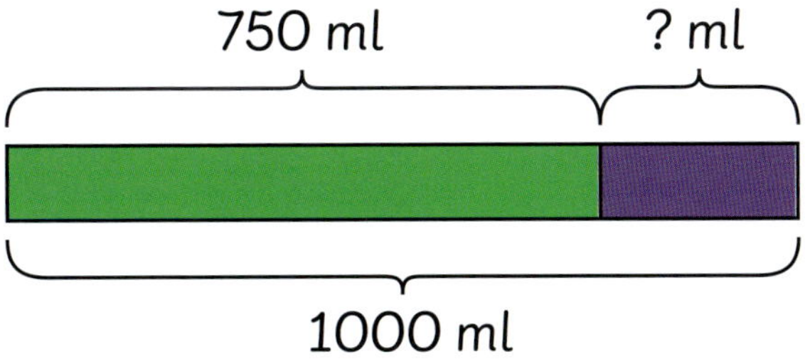

1000 – 750 =

Another ml of water is needed to make 1 l.

Complete Worksheet 5: pages 195–196

Writing Capacity in Litres and Millilitres

Lesson 6

Explore

I tried to pour 2 l of water into my bottle but I am left with this.

What is the capacity of Lulu's water bottle?

Master

The bottle took 1 whole litre of water and part of a second litre.

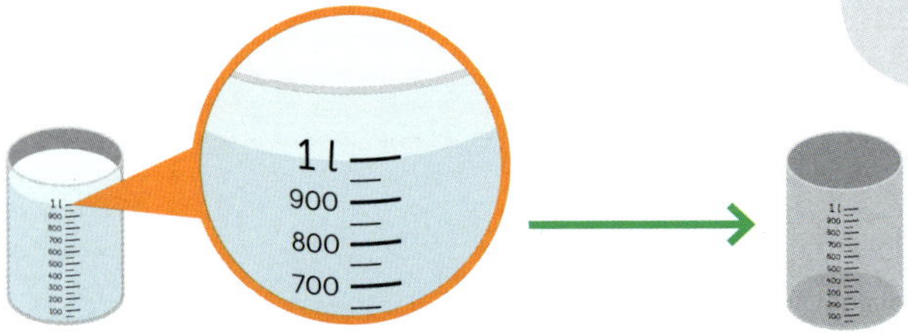

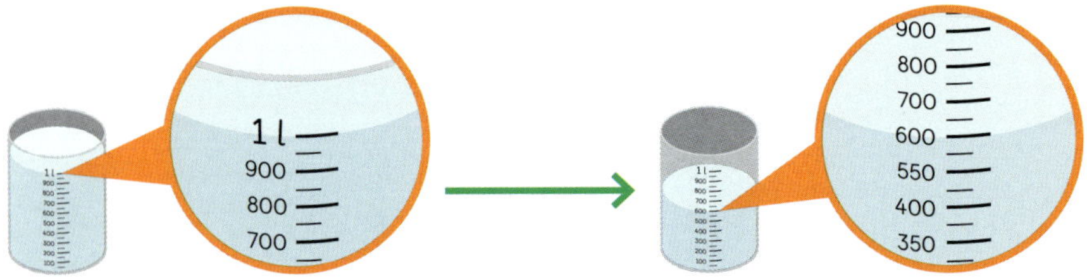

There is 600 ml left in the second beaker.
1000 – 600 = 400

The water bottle took 400 ml from the second beaker.

The capacity of Lulu's water bottle is 1 l 400 ml.

Activity Time

Group work (4 pupils)

You will need:

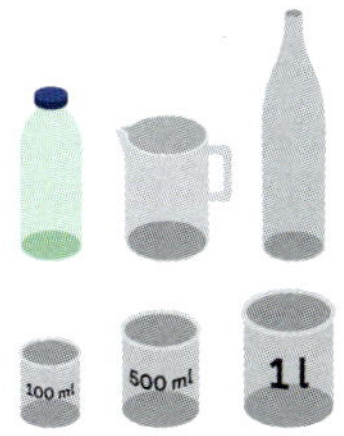

1. Guess whether the capacity of each container is more than 1 l, less than 1 l or about 1 l.
2. Fill each container with water, then pour the water into  to measure its capacity.
3. Record your results in a table.

Container	More than 1 l	Less than 1 l	About 1 l	Check
plastic bottle	✓			1 l 500 ml

Guided Practice

1 Water from two filled 1-l beakers is poured into a jug. The amount of water left is shown.

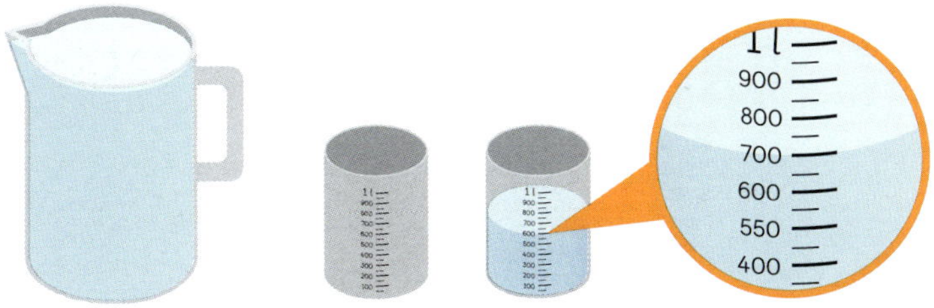

The capacity of the jug is ______ l ______ ml.

2 How much water can the watering can hold?

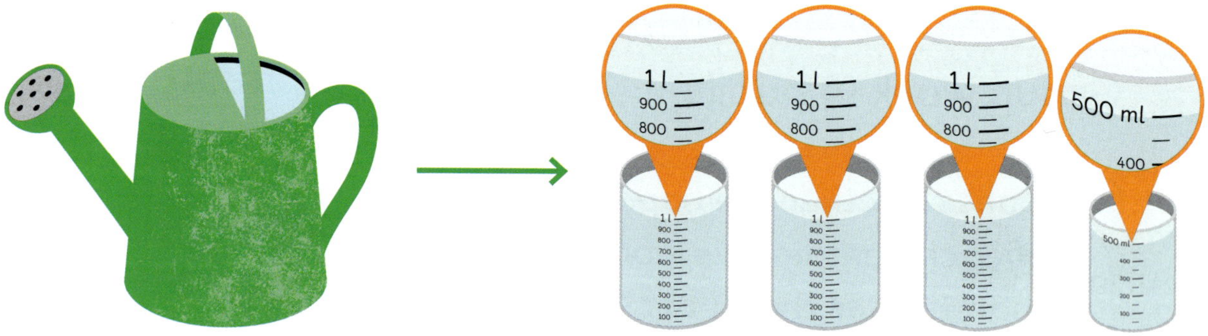

The capacity of the watering can is ☐ l ☐ ml.

3 What is the capacity of a can of paint?

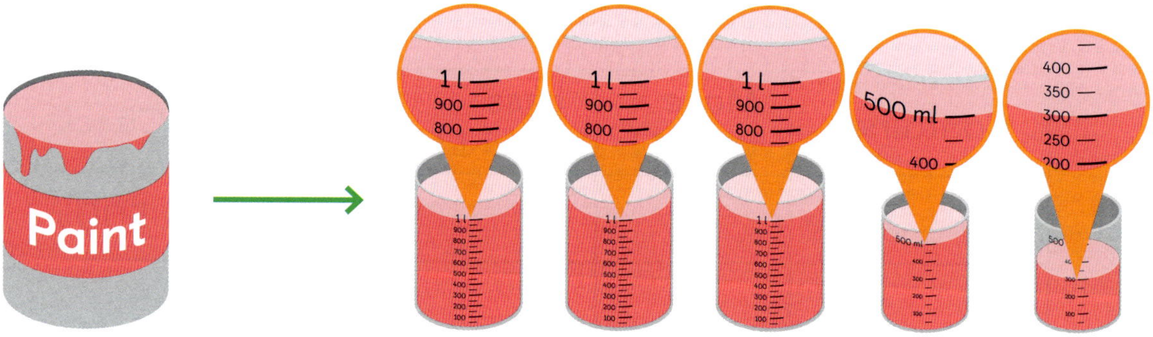

The capacity of the paint can is ☐ l ☐ ml.

4 How much water can the vase hold?

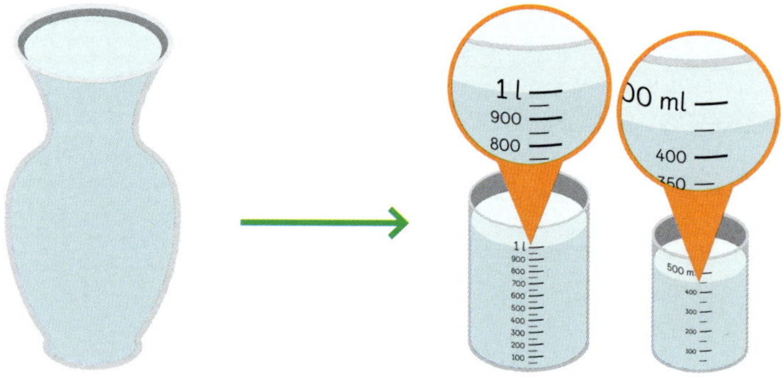

The capacity of the vase is ☐ l ☐ ml.

Complete Worksheet **6**: pages **197–198**

Solving Word Problems

Lesson 7

Explore

Emma's water bottle is full. It has 175 ml less water in it than in the beaker. What is the capacity of Emma's water bottle?

Master

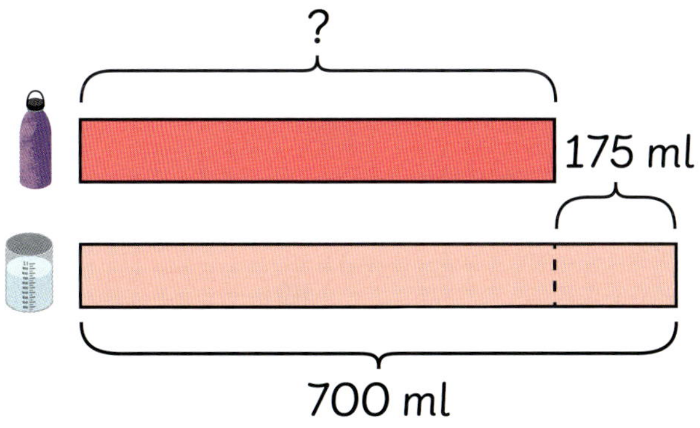

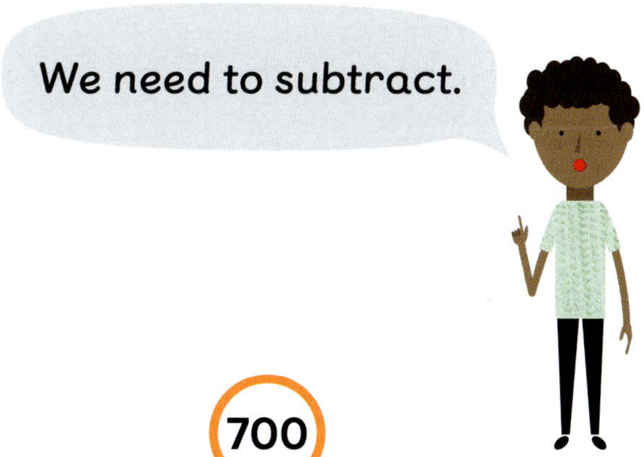

	7	0	0
−	1	7	5
	☐	☐	☐

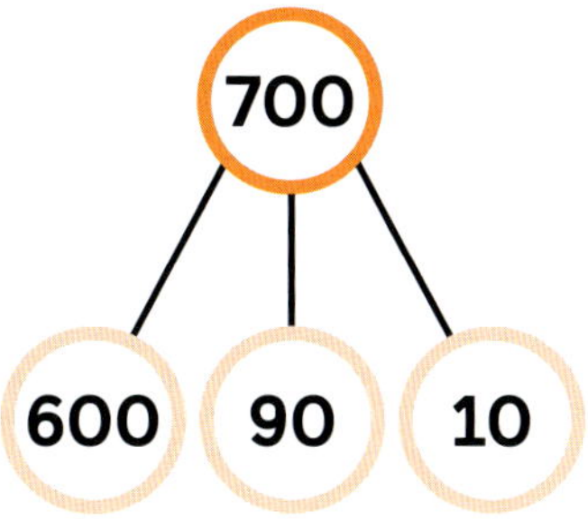

700 − 175 = 525

The capacity of Emma's water bottle is 525 ml.

Guided Practice

Solve the word problems.

1 The capacity of a small mug is 250 ml.
Its capacity is 115 ml less than the capacity of a larger mug.
What is the capacity of the larger mug?

2 The volume of water in the clear vase is 700 ml.
This volume is 345 ml more than the volume of water in the purple vase.
Find the volume of water in the purple vase.

3 The volume of water in the kettle is 60 ml more than the volume of water in the 350-ml mug.
Find the volume of water in the kettle.

Complete Worksheet **7**: pages **199–200**

Solving Word Problems

Lesson 8

Explore

The volume of water Jacob's family used on Monday was about 290 l. It was about 65 l less than the volume of water they used on Tuesday. About what volume of water did they use over the two days?

Master

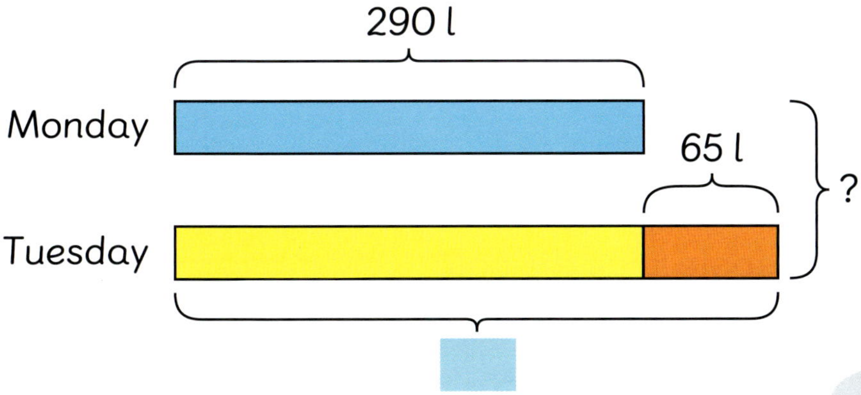


```
    2  9  0
+   3  5  5
-----------
    □  □  □
```

290 + 65 = 355

290 + 355 = 645

The volume of water they used over the two days was about 645 l.

Guided Practice

Solve the word problems.

1 A paddling pool has 437 l of water in it. A smaller paddling pool has 78 l less water in it than the larger paddling pool. Find the total volume of water in both paddling pools.

2 Oak made a smoothie using 260 ml of orange juice and some apple juice.
She used 190 ml more apple juice than orange juice.
What was the total volume of juice Oak used?

3 A café used 178 l of apple juice in September.
It used 39 l less apple juice in October.
What was the total volume of apple juice the café used over the two months?

Complete Worksheet **8**: pages **201–202**

Solving Word Problems

Lesson 9

Explore

Lulu's bottles hold 4 l of water each.
Lulu has twice as much water as Elliott has.
How much water does Elliott have?

Master

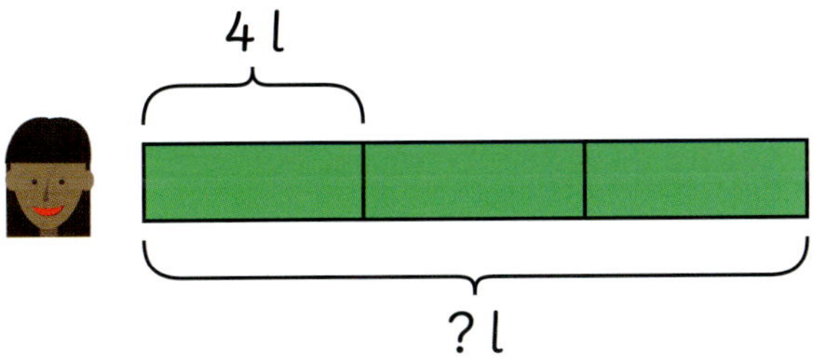

$3 \times 4 = 12$
Lulu has 12 l of water.

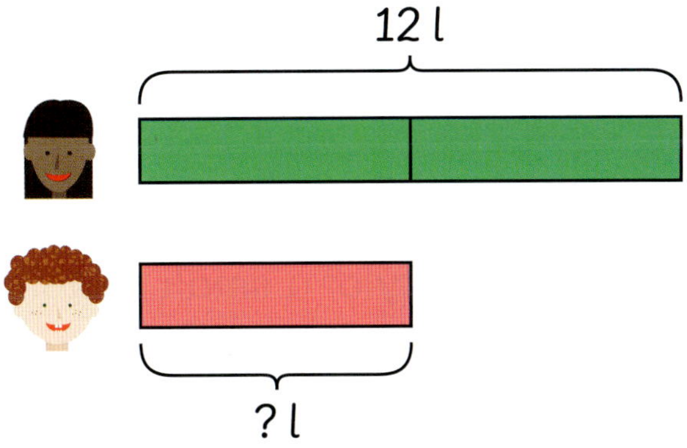

$12 \div 2 = 6$
Elliott has 6 l of water.

Guided Practice

Solve the word problems.

1. The capacity of a large glass is twice the capacity of a small glass.
The capacity of the small glass is 375 ml.
What is the capacity of the large glass?

2 A recipe for baking a loaf of bread uses 12 ml of vegetable oil for 1 loaf. How much vegetable oil is needed to bake 4 loaves of the bread?

3 The large tub has 4 times the capacity of the small tub.
The capacity of the large tub is 36 l.
What is the capacity of the small tub?

4 The capacity of a small paddling pool is 96 l. The paddling pool has 8 times the capacity of a bucket.
What is the total capacity of the paddling pool and the bucket altogether?

Complete Worksheet **9**: pages **203–204**

Solving Word Problems

Lesson 10

Explore

Each fuel container can hold 27 l of fuel.
After filling 8 fuel containers, there is 68 l of fuel left in the drum.
What was the volume of fuel in the drum before the fuel containers were filled?

Master

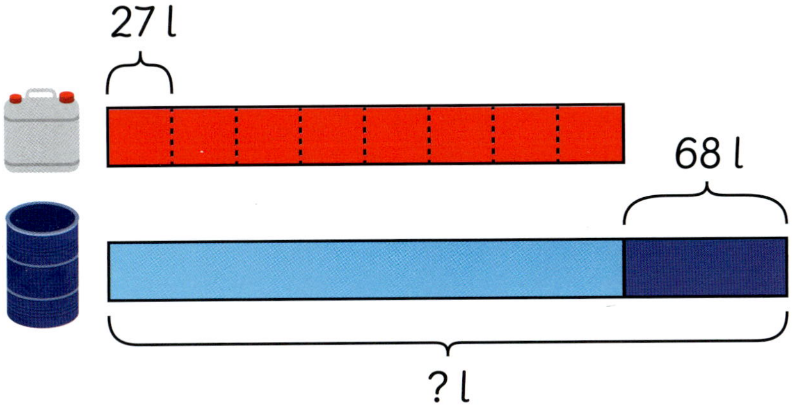

$$\begin{array}{rrr} & {}^{5}2 & 7 \\ \times & & 8 \\ \hline 2 & 1 & 6 \\ \hline \end{array}$$

27 × 8 = 216
216 + 68 = 284

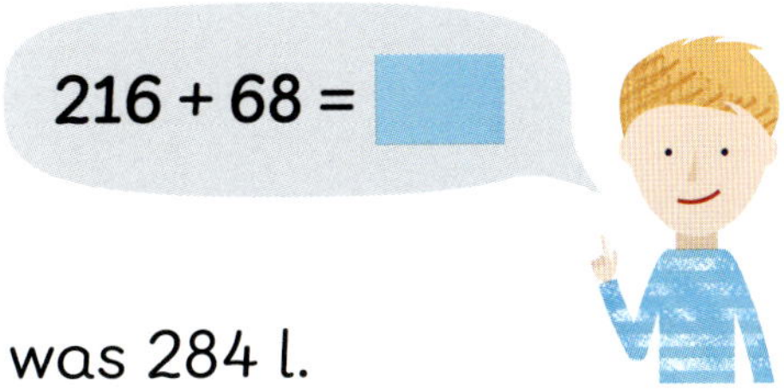

The volume of fuel in the drum was 284 l.

Guided Practice

Solve.

1 There is 3 times as much water in  as there is in .

The total volume of water in both containers is 16 l.

What is the volume of water in  ?

2 After Amira fills 5 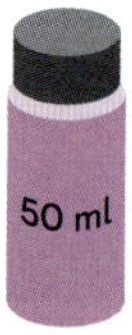

with liquid soap, there is still 684 ml of soap left over.

What was the volume of soap she had at the beginning?

3 Amira thinks she can fill 3 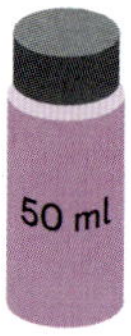

using 180 ml of soap.

Is she correct?

Complete Worksheet 10: pages 205–206

Mind Challenge

carton A carton B carton C

The total volume of milk in the three cartons is 500 ml.
The volume of milk in carton A is 70 ml less than the volume of milk in carton B.
The volume of milk in carton C is half of the total volume of milk in the 3 cartons.
Find the volume of milk in each carton.

My Maths Journal

Look at the different containers your teacher has provided.
Guess if the volume of liquid in each container is greater than, less than or about 1 l.
Check the volume of the liquid using a measuring jug.
Use a table to record your results.
Share your results with your class.

Container	More than 1 l	Less than 1 l	About 1 l	Check
red bowl		✓		650 ml

Review

I can...

- [ ] measure volume in millilitres (ml) and litres (l).
- [ ] measure capacity in ml and l.
- [ ] write volume in ml and l.
- [ ] write capacity in ml and l.
- [ ] estimate and check a volume of liquid.
- [ ] estimate and check the capacity of a container.
- [ ] solve word problems involving volume and capacity.